LET'S MAKE IT
FROM JUNK

LET'S MAKE IT FROM JUNK

By Eileen Mercer

Photographs by Ed Bievenour

Drawings by Kevin Franklin

Stackpole Books

LET'S MAKE IT FROM JUNK

Copyright © 1976 by
The Stackpole Company

Published by
STACKPOLE BOOKS
Cameron and Kelker Streets
Harrisburg, Pa. 17105

Printed in the U.S.A.

Library of Congress Cataloging in Publication Data

Mercer, Eileen.
  Let's make it from junk.

  1. Handicraft. 2. Waste products, I. Bievenour, Ed. II. Franklin, Kevin,
III. Title.
TT157.M46   745.5   76-3551
ISBN 0-8117-0939-6

This book is dedicated
to my daughter,
Wileen,
a faithful source
of throwaways and castoffs

Contents

Acknowledgments   11

Introduction   13

Dimensional Painting   15

Snap-on Curler Napkin Rings   17

Bubble Pictures and Photos   19

Beverage Can Beauty   20

Tape Dispenser Candlesticks   22

Bird Cages and Cups   23

Offbeat Hanging Baskets   25

Miniature Spoon Landscapes and Fork Holders   27

Needlecraft Yarn Palette   29

Earrings on a Trimming Wheel   31

Button Bouquets   32

CONTENTS

Jewelry Potpourri 33

Kitchen Cleanser Containers 35

Bookmarks 37

Metal Spout Butterflies 38

Glass Dome Displays 41

Plastic Mold Displays 43

Standing Bead Mobile 45

Snake Charmers 46

Fooling with Aluminum Foil 48

Giant Paper Clip 51

Trinket Boxes 54

Uncluttering Kitchen Clutter 55

Stationery with Iron-on Silhouettes 60

Bulbous Flower or Sea Creature 62

Easels for Many Uses 64

Patterned Aluminum Jewelry 66

How Sweet It Is — a Lemon! 68

Imitation Shell 70

Jeweled Pigtails 73

Tin Lid Pendants 76

Tinsel Bouquet 78

Old Belt Sewing Kit 80

Stamp Dispenser 82

Strike Up a Match! 83

"Empty-headed" Pincushion 85

Eyeglass Case—a "Spectacle" Plus a Purse 86

[8]

CONTENTS

Flash-Cube Belt and Jewelry   88

Easy Crazy Quilt   89

Ecology Boxes and Pendants   93

Second Life for Candle Containers   96

Button Portraits   98

Crazy Helping Hand   100

Dictabelt Hat and Belt   102

Dolly's Clothes Hangers   104

Candle in a Bottle   106

Wallpaper Patchwork   108

Whimsical Pull-off Toppers   109

Canned Hang-ups   111

Cash Cache   113

Planter-Tray Compartments   115

Heavy Jug Book Ends   116

Foil Pie-Tin Centerpiece   118

Margarine-Container Flowers   120

Clown Bank   122

Mason Jar Lid Pendants   124

Paint Applicator Desk Accessory and Easel   125

Old-fashioned Clothespins   126

String or Twine Holders   128

Styrofoam Packing   130

Meat-Tray Galleries   136

Pants-Hanger Candlestick   138

Postage and Pill Boxes   139

[9]

CONTENTS

Personal Bulletin Board   140

Papier-mâché Projects   142

Sewing Kits   145

Vases from Cosmetic Containers   148

They're Not Ice Cream Cones!   149

Who Sits Where?   153

Final Throwaways and Castoffs   155

Index   159

Acknowledgments

The following chapters are reprinted with the permission of *Popular Handicraft & Hobbies:*

Ecology Boxes and Pendants

Crazy Helping Hand

Sewing Kits

Uncluttering Kitchen Clutter

Jeweled Pigtails

Introduction

My first book, *Let's Make Doll Furniture,* described how to make doll furniture from throwaways and castoffs. This hobby consumed my attention for some time.

Later I branched out into other areas. Now there is scarcely anything I won't attempt to recycle to give it a second chance. Relatives and friends sometimes expect next to the impossible, but they keep me alert and are faithful sources for dumping things into my junk bins.

You might not have the same kind of junk I have or acquire. On the other hand, you may have something better from which to create practical things or *objets d'art* and thus grant second life to the original item.

So in most of the projects in this book you are given options. Use other material or another object that serves equally well if you don't have what is suggested. This approach enables you to switch on the light bulbs of your own imagination to make your handiwork distinctive and highly personal.

Before beginning a project, read all the instructions and suggestions carefully. Alternatives may come to mind as you proceed.

I consider making something useful or decorative out of practically nothing a genuine craft, worthy of recognition.

Have fun turning your junk into treasures!

Dimensional Painting

The "painting" in the photograph is an outstanding example of something really attractive that can be made from practically nothing—junk and throwaways.

If you can't conjure up something to "paint" in your mind's eyes, look through magazines, children's coloring books, et cetera. Select a subject that has simple lines—parts that can be suggested by putting together small items. A train or an automobile offer good possibilities, as well as a house or a barn landscape, with trees and clouds. It is not necessary to show all details—just enough to suggest what the subject is.

The lighthouse scene in this photograph is mounted on red velvet stretched over a 16″ x 20″ artist's canvas board. The cloth background can be upholstery material, burlap, terry cloth—almost anything that suits your fancy and the scene selected. But don't choose a background that is too elaborate. It should be a solid color, or perhaps a small tweed, so as not to detract from the scene. The cloth is glued or taped securely to the solid background piece, which can be masonite or plywood, even heavy cardboard if the picture is not too large.

This lighthouse scene is done in black and silver. It was not necessary to paint some of the objects. Be extremely careful not to smear glue where it is not supposed to be. Put glue on the object and let it get tacky before placing the object on the background. But first set up your entire scene in a practice run before you begin gluing anything in place. This means, of course, that all your junk pieces are painted and thoroughly dry.

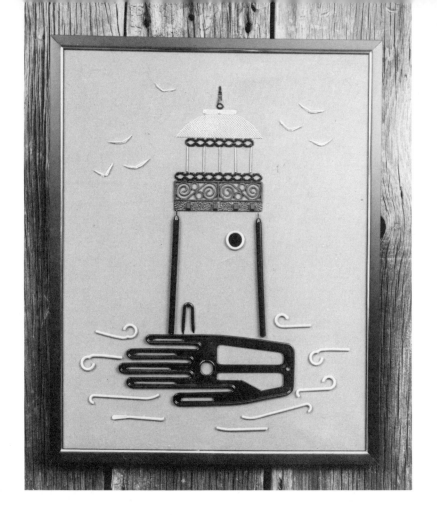

Here is a description of the various items in the lighthouse scene, beginning at the bottom: Waves are plastic clips sometimes attached to pieces of new clothing, like socks. They are broken apart, painted silver.

The ground/rock foundation is a plastic shape for stretching and drying gloves. It is painted black.

Sides of the lighthouse: Two 6"-long pens painted black. Door: Large staple painted black. Window: Plastic curtain ring painted silver. The "glass" in the window is a piece of film (black) cut in a circle and glued to the ring.

Walkway/railing above the two pen sides: Black plastic hanger that held four measuring spoons.

Above the walkway/railing: Five 1¼" finishing nails.

Over the nails: A 3½″ length of fancy popsicle stick painted black.

Dome: Cut from a scrap of screen discarded from an old radio.

The 1¾″ strip over the dome: Cut from the silver cutting edge of a box of foil.

Pointed spire: Metal hanging piece from a clothlike picture hanger, painted black.

The flying birds are cut from the rolled edge of a foil pie tin, the pieces slightly pushed apart to be shaped as wings.

Frame your work of art. Then sit back and admire it!

Snap-On Curler Napkin Rings

Probably every household has hair-curling gadgets no longer in use. They usually turn up in quantities at rummage and garage sales. If you do not have any discarded snap-on curlers, undoubtedly you could find some at one of the thrift sales.

I "inherited" a bag of snap-on curlers in some junk that was given to me. Out of the snap-on clamp that goes over the curler evolved an attractive napkin ring. A clamp approximately 2½″ long would be about 1½″ diameter if it were a complete circle, and this is what I used.

The open spaces, or "long holes," are filled in with yarn. You can use matching or contrasting yarn. Thread a yarn needle (blunt point) with a piece of yarn about 2½ yards long. This is probably more yarn than you will need, but it is better to have some left over than not enough to complete a ring.

Holding the open part of the clamp toward you, upright, tie the end of the yarn twice on the right side at the bottom opening. Turn the knot to the inside. Put your needle under the first bar, where you just tied the knot; bring the thread up and over to cover the knot. Go to the next vertical bar; take a back stitch through the second bar. Bring the needle up to carry the yarn on top of the back stitch. (Every time you take a stitch, keep the yarn at the top of the little bar.) Continue around the curler, completing the first row.

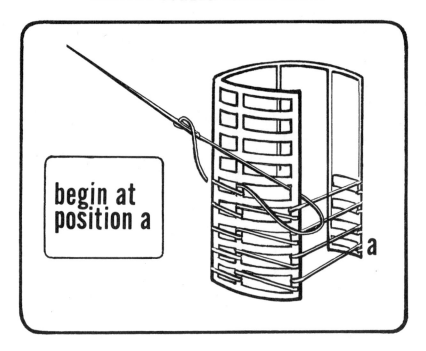

begin at
position a

After you have made the back stitch on the last bar (left-hand side), carry the yarn across the opening to the same place you started, making another back stitch. Then go up to the second vertical bar (on the right-hand side) with a back stitch. Work across the entire second row, as you did the first. At the open space always cross over to the row where you began; then go up to the next row on the starting side (right hand). This keeps the bars or rows of yarn straight across.

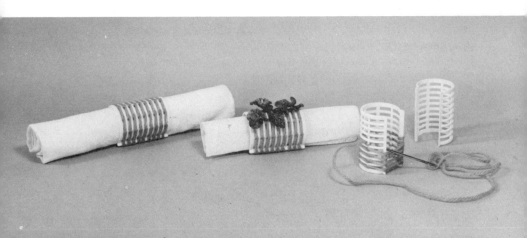

When you have finished at the last row (right-hand side), make two or three back stitches in the yarn on the inside. If necessary, while working in this cramped space, you can bring needle and yarn out at the opposite end.

The napkin holder rests on the flat surface made by the rows of yarn crossing the space that was open.

Add some small flowers for further attractiveness. The sample shows clothlike flowers from a discarded hat.

Bubble Pictures And Photos

Many of today's products are packaged on cardboard or in boxes with contoured clear plastic "bubbles" covering them. Some of these bubbles are rather flimsy; others are strong and durable.

You can use these better bubbles to cover arrangements of flowers and pictures. When you have a new bubble, remove it carefully so as not to dent or tear it.

Select a compatible background for what you wish to mount. In the largest picture in the photograph, a piece of cardboard is covered with green crushed velvet. The edges of the material can be glued to the back of the cardboard, fastened securely with masking tape, or you can crisscross back and forth with needle and thread to hold the material in place.

Under this bubble there are two small plastic lavender iris, a purple flower, and some greenery, together with a jeweled, gossamer butterfly. Gold braid is glued around the edges of the bubble after it has been first glued to the background. The frame is gold.

The smaller flower arrangement is fitted into a white plastic frame. The background is a piece of blue satin ribbon. Two sprigs of pearlized buds or flowers, with blue cloth leaves, are combined with dried brown leaves and a sweet-gum ball, all glued in place.

The bubble over the photographs was double, so I used it to encase pictures of my two youngest grandchildren. To set it apart as a little more special than just a photograph, I glued some dried brown leaves and pods on the front of the frame.

Beverage Can Beauty

Almost everyone, at one time or another, has probably linked together the pull-off openers of beverage cans. These creations can look like rubbish, or they can, with a little care, be made into things of beauty and usefulness.

All the pieces should be the same kind, new, and handled with care, so that they are not unduly bent or smashed. Watch out for the sharp edges of the openers. Wearing gloves will help avoid the danger of cuts.

Here is an attractive belt which will brighten up a costume. What takes it out of an ordinary class is the addition of yarn fringe.

You can use one color of yarn, or a combination. The sample has red-black-red looped on each ring. The "knot" part of the fringe

is the top side. The length and thickness of the fringe can vary according to your taste, but don't be skimpy. When all the fringe has been affixed, trim ends neatly the same length.

To fasten the belt when you are wearing it, just link it together. Or, a crocheted or macramé-type closure for tying could be made from lengths of the yarn.

The floral picture is also made from these pull-off beverage can openers. It is mounted on an artist's canvas board painted black. The frame is silver. Other types of background could be used, such as burlap, upholstery material, velvet, et cetera.

To use an entire ring, clip it apart where the narrow band and its base are joined. Twist into an S curlicue. Bend some both ways so that each will be right side up.

Use the narrow part of the ring only, clipped away from its base. Also, use the thicker base part by itself. Use the entire ring, separated from the "petal" part; also, use the petals themselves.

To fringe a petal, cut the rounded part into narrow slivers, letting them curl at will.

To make a large flower of single petals, cut a circle of cardboard a little smaller than you want the completed flower. Space evenly a row of petals around the outside rim of the cardboard. Practice placement before gluing the petals in place. Add two or more layers of petals. If necessary, clip away part of the inner row before gluing on a center. The large flower in the photograph was made with the gold side of the petals showing. The remaining flowers are silver. Any, or all, of this flower material could be painted, if you wish.

Centers are buttons or flat pieces of discarded earrings or other jewelry. The foundations for self-covered buttons make excellent centers.

From the sample photograph you can get ideas as to various shapes and combinations that you can create.

Besides a basket arrangement, you could have stemmed flowers. Cut stems from foil pans or the metal tear-off rims of foil or waxed paper.

Practice laying out an arrangement before starting to glue anything into place.

Tape Dispenser Candlesticks

Sometimes it seems as if we couldn't survive without modern, dependable self-sticking *Tape!* When the tape's all gone, throw away the little dispenser? Never! These castoffs make perfect candlestick arrangements.

In the photograph you will see a clear plastic holder with a yellow candle, a sprig of plastic greenery, and a couple of ceramic figurines. There is a circular opening shaped into the side of the dispenser (now the bottom), and the lady fits perfectly into this "hole." The man stands behind her. The lady's dress is yellow, which blends with the yellow candlestick.

Several dispensers can be put together to form various patterns or arrangements. They need not be glued, but merely placed to-

gether in the manner desired. The height of the candles can vary, or they can all be the same length. The candles can match the color of the holders, or they can contrast. The arrangement in the photograph uses pink holders, candles, and flowers.

Bird Cages And Cups

The bird has flown and the cage is empty! But you won't have to stash away the cage. Instead, pretty it up and perk up a dull spot in the house.

The cage in the photograph is a mini cage, usually used for transporting a bird. The same idea can be used with a larger, regular-size cage.

Clean and repaint any parts that need it. You can also add color to the perches, if you like.

The sample cage, with an orange bottom or tray, has flat, fernlike pieces of orange plastic placed in the bottom. A large sprig

of greenery and orange flowers highlights one corner, and several blossoms are affixed to the perch. These flowers hide the feet of the bird, which have been wired to the perch so that it remains upright. The bird has colorings of yellow, brown, white, and black.

Add a bow at the handle, hang up the cage, and see a dull spot sing!

Then look what you can do with little plastic water and feed cups.

The hanging cup: With a heated pick or awl, punch a small hole in the top of the cup. Use fine wire, small cord, or several strands of thread to make a hanger. The sample has loops of blue thread, pushed through the hole with a needle. Tie a knot in the ends of the thread. The knot pulls up against the hole on the inside, with the loops on the outside for hanging.

Fitted into the cup are dried weeds and flowers, together with a small length of green/blue tinsel wadded in the bottom. Half a dried flower bud is glued on the outside protrusion of the cup, and, over this, a blue glass bird.

Easel-type cup: This cup has three red pushpins—two in front, one in the back—to make standing feet. Hold the cup firmly and wiggle the pins back and forth as you push them inside, so that the plastic will not split or crack. White plastic lilies of the valley are stuffed inside, and a red-hatted, red-costumed miniature singing figure looks out the "window." A small red velvet heart, edged in simulated pearls, is glued on the front protrusion.

Offbeat Hanging Baskets

Hanging pots and baskets for plants and flowers are both practical and decorative. They are extremely popular. You see them almost everywhere. Let's hang up something that is eye-catching and a bit out of the ordinary.

How about starting with an old hat? The Mexican-type hat in the photograph has a frazzled or fringed edge. There are many other types of straw hats, in different sizes and shapes. Ladies' or men's hats are suitable. If the hat is too plain to suit your fancy, glue or sew on braids, cords, shells, beads, or other decorations.

The smaller hat in the photograph originally belonged to a large doll. The open wickerwork basket was not a hat, but a be-ribboned, beflowered Christmas bell that had outlived its usefulness.

Place the hat in a rope or cord hanger—one you make yourself or one you can purchase. If you're working with macramé, you've probably already made hangers. Or, a hanger is easily assembled from eight lengths of rope or cord, merely by tying simple knots.

Put the potted vine or plant in the hat, and that's all there is to it. Step back and admire what you've created. Isn't it pretty?

Another castoff that makes a good hanging basket or vase is an old light-fixture shade. If the shade has a rim or ledged edge

where the screws held it to the electrical fixture, you can use chain for hanging.

You will need lightweight brass or gold-finished chain. Use pliers to separate and rejoin the links. Fasten a length of chain around the globe, under the ledged rim.

Separate three more equal lengths of chain, from 12" to 15" each, according to the size shade you have and the place where it will hang. Fasten the three lengths of chain equidistant into the chain around the shade. Then fasten the ends of the three lengths of chain onto a ring or hook for hanging. A metal pull-off top from a beverage can makes a suitable hanging ring, as does a metal or plastic curtain ring.

A small shade bowl could be filled with dirt to grow live plants or flowers, but a larger one might be too heavy when it's holding dirt. You can fill the shade with plastic flowers and greenery or with dried weeds and flowers. Use fresh flowers, if you wish, in a container of water set inside the shade.

If you have an old shade that has no rim, use a rope or cord hanger, as for the hat.

Miniature Spoon Landscapes
and
Fork Holders

If you have odd spoons rattling around in a drawer, you can turn them into adorable miniature landscapes. Perhaps you have been saving a baby spoon, from this you could create a real keepsake. These small arrangements are ideal as remembrances for someone in the hospital or a convalescent home. They take up little space but are very attractive.

The samples shown will give you ideas as to how to bend the handle of the spoon. Protect the finish of the spoon by using cloth or something that will not mar or scratch. If you cannot bend the spoon by hand only, hold it over the edge of a counter and use a broomstick or something similar to brace it.

When you have a pleasing shape and one that will hold the spoon upright, you are ready to proceed with the decorative landscape. Materials used can be dried weeds and straw flowers, plastic greenery and flowers, aquarium gravel, seeds, ceramic figures. These are all anchored with bits of papier-mâché or glue or a clay that will harden. Always try to cover up or camouflage the base where objects are anchored, so that the entire arrangement looks professional, not amateurish.

Here is a description of each of the spoons.

Pair of deer: White aquarium gravel is glued in the base of the spoon, into which are also glued bits of small plastic greenery and cut-off pieces of a spire flower. Two miniature deer are glued on top of the rocks.

Bird on nest: A dried flower is anchored in clay in the bowl of the spoon, simulating a nest, on which a feathered and flocked bird sits. Moss and weeds and strawflowers are also stuck into the clay.

Horse: Green aquarium rock is pushed into florists' clay (a few pieces that would not stick were glued here and there), as well as a small thistle and some dried weeds. Tweezers are handy to use to pick up the aquarium rock. A miniature horse is glued on the rocks.

Toadstool: Moss, dried weeds, strawflowers, and a ceramic toadstool are pushed into florist's clay. In front a few dried cantaloupe seeds are glued on the spoon bowl.

Besides "landscaping" teaspoons, you might like to try a soup or dessert spoon, as well as the larger tablespoon. If you have no odd pieces of silverware, you can always pick up inexpensive pieces at garage or rummage sales.

FORK HOLDERS

I found some odd forks also jangling around in a kitchen drawer. What to do with them?

Well, bend up the tines of the fork, letting the handle lie flat. This makes an excellent holder for a recipe card while you're cooking or baking.

If you would like to pass on your favorite recipe to a friend, it can be typed on a 3" x 5" card. You can add a brief pertinent message; your name can be written or printed with a felt pen in colors. Add a flower decoration cut from contact paper, or a motif clipped from a gold paper doily.

Salad forks, with shorter handles, can also be used, as well as small ice cream and cocktail forks.

Another sample fork has a small plastic flower and greenery twined through the tines. It holds a get-well verse that was salvaged from a used greeting card. Cut out a verse to approximate size; then "antique" it, which is done by burning the edges. If you have never done this burning over a candle, or other flame, practice first on a card other than the one you want to use. Hold the card vertical by the paper and let catch fire for an inch or two; then quickly blow out the flame. Continue in this way around the card edges. Then break away the portions that are heavily charred, leaving a nicely browned edge. This is a thoughtful remembrance for someone in the hospital, or you may want to give it to someone in your own home.

The third fork holds a burned-around-the-edges message that says "MISSING 'U.'" Two flowers cut from contact paper decorate the card, which was also salvaged from a used greeting card.

Some people read and/or memorize daily a printed Bible verse. One of these forks would make a very nice holder for this commendable habit.

Put an appropriate message on the breakfast table, a tray, or in a noticeable spot for your favorite person. Such a gesture will brighten up anyone's day. Change the message from time to time for that special someone.

Needlecraft Yarn Palette

What is more frustrating than trying to untangle scads of yarn or thread that was not organized at the beginning of a creative stitchery, crewel embroidery, or needlepoint project?

Save your nerves (as well as supplies) by making a serviceable palette or holder for those cut lengths of yarn.

A simple, satisfactory holder can be made from the rim of a plastic lid. Use whatever size is needed or desired. Simply cut away the flat inner part of the lid and use the rim only. Try using pointed scissors or a sharp knife to do the cutting. There should be no notches or rough edges remaining.

Loop the various colors over the rim, then pull out one

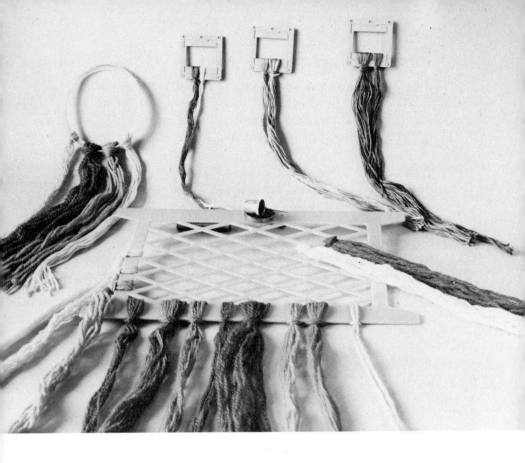

strand at a time by picking up the horizontal bar or "knot" of the loop.

If you have some frames for housing slides (they are about 2" square), you have a handy gadget to recycle. Loop threads over one or over all four sides of the frame for plenty of storage space.

A larger palette is a plastic latticelike lace winder which are discarded by the dozens in department and variety stores. You can usually have them merely for the asking. At least, I have always found this to be so. Then you will have bars galore over which to loop yarns.

Also, at the top of the winder there is space to glue on a small magnet for holding needles while you are working. If the magnet is large enough, it will even grab hold of a thimble. The magnet in the sample is part of a rusted hook magnet. The hook was broken away.

I hope your yarns get straightened out in less time than it did to do mine!

Earrings On A Trimming Wheel

In the last chapter a lace winder was used for making a yarn palette. For an earring holder or stand, you will need a trimming reel, which is another plastic gadget available from department and variety stores where braids and fancy trims are sold. You can undoubtedly have these trimming reels, also, for the asking.

There isn't anything to do except cart it home and remove any stickers or labels affixed. Obstinate gooey adhesives can be soaked away with cooking oil. Sometimes I use a little turpentine or paint thinner in the final stages of removal. But using one of

these liquids must be done lightly and quickly, as a plastic finish can be marred by some chemicals.

If you do not want the holder to stand up, it can be laid flat in a drawer. It could be painted or decorated in some manner, but is attractive enough just as it is. I "tumbled" my first name down one side, with self-sticking letters, just to give you an idea. A gold self-sticking seal also looked nice.

Button Bouquets

For years bouquets have been made from buttons. I have seen some that were far from pleasing or attractive—merely bare wire stuck through the two holes of white buttons. That is a *bouquet?*

If you want to make a bouquet from the proper kind of colored buttons, you will find it exceptionally beautiful.

Use only shank-type buttons—those with no holes showing on top. Exceptions can be made with really elegant decorator buttons, but in the holes insert tiny seedlike stamens used in making artificial flowers. The stems are fastened on the under-

side and are covered by the green calyx. Also, tiny jeweled buttons may be glued over, or into, large holes of a large button, or into a cuplike button. Buttons can be flat, concave, convex, or square. Gold and silver and jeweled buttons are especially attractive.

I like to use a lightweight, plain green metal wire for stems, as it looks delicate. However, other types of wire could be used, and it could be covered with floral tape. Put a length of wire through the shank holes, twist together securely against the bottom of the button. Then add a flower calyx (obtainable at craft and hobby stores) of suitable size to fit the button. This completes the flower in a professional-looking manner and lends airiness to the otherwise staid bouquet.

The calyx may be glued or affixed with a bit of any kind of clay that hardens, according to the shape of calyx used (flat or cupped), as well as the type button. However, I usually do not fasten the calyx except by pushing it very tightly against the button and then making an almost imperceptible bend in the wire with small, long-nosed pliers. This holds everything together securely.

The bouquet in the small blue/silver vase is made from four large blue buttons with holes. The stem wire was put through both holes and twisted. Then a flat green button (with shank removed) was glued on top. A bit of plastic greenery is used, as well as many pieces of fine silver wire, shaped into sweeping curves. The lengths of wire can be held together in one hand, curved into shape with the other, then separated as needed when you are placing it in a vase.

Jewelry Potpourri

Wearing a piece of jewelry you have created yourself can be rewarding—you get talked about!

The large necklace and earring set in the photograph is not for a petite or diminutive person, as it might be a bit overpowering. But many larger people could wear it well.

The loops are the push-together rings used for hanging a shower curtain. The ones in the sample are black.

Links of an old chain were separated and used to hold the loops to the neck piece. Use a link and a jewelry jump ring to suspend the earring loop, to which is attached whatever type earring you wear—pierced or clip. Jewelry findings are obtainable at hobby and craft stores.

I doubt if you could ever guess what the other necklace is made from. My granddaughter is diabetic and uses insulin. She gives me the little white plastic caps (round pieces) and another red gadget (oblong pieces) from her insulin supplies.

You might not have access to these same items, but perhaps something else would do equally well, for most plastics of this nature melt satisfactorily.

Lay the pieces on a large foil-covered tray so that they do not touch one another. Put them in an oven heated to 350° to 375° to melt them down. Watch closely if you are not accustomed to working with this type project.

When the plastic has melted or softened sufficiently, you can pull the piece of foil into the sink and run cold water over it. Be careful not to burn your fingers!

After the pieces are cool, use a heated pick or awl to punch holes through which you affix jewelry jump rings to assemble in whatever pattern you choose.

If you want dangles, the piece to which it is attached will need an additional hole (besides the two for jump rings) a little to one side of the bottom hole. For left and right sides, remember to make the holes accordingly—on opposite sides of the bottom hole.

Fasten the necklace with a closure from an old piece of jewelry if you happen to have one.

Earrings can be made to match.

Kitchen Cleanser Containers

One of the most common, necessary products in all homes is the scouring cleanser used in kitchens and bathrooms. The empty cans make excellent containers for many things.

For any of the uses described, weight can be added to the container by putting in a small quantity of clean, dry sand or aquarium gravel, if this seems necessary.

If the holes in the lid are not large enough, they can be made larger by reaming them out with a large screw. Or, a large nail may widen them sufficiently. Additional holes may also be punctured in the lid.

Decorate the can according to its intended use.

I have been a firm believer that a sprinkle of sugar in cooked vegetables enhances the flavor. For years I used a handy cleanser can to hold sugar for this purpose. The can has a printed SUGAR label. (Of course, a container must be thoroughly cleaned before being used for food.)

One of the smaller cans makes an excellent holder for crochet hooks. The four oval panels that contained advertising lettering were covered with pieces of felt cut with a pinking

shears—two dark blue, one light pink, one rose. The can is yellow.

A taller orange-colored can holds knitting needles. This is one of the newer-type cans, with the sliding lid, which has two sizes of holes. It is decorated with three narrow strips of harmonizing contact paper, affixed around the circumference, plus a flower/leaf motif on the front.

A white can, banded with a strip of green/blue/white/black contact paper, becomes a vase to hold flowers made on a daisy wheel from *soutache* braid. (The making of these flowers is another project.) The colors of the flowers match those in the contact paper trim. The container can hold water and living flowers as well. When the youngsters bring in those dandelions and other beautiful weeds, one of these containers makes a serviceable vase that won't break. Just put the stems through the holes in the vase.

Another can to hold stick incense has been sprayed silver.

The bottom portion is banded with a gold strip of thin plastic material that was originally a purse. A flower is cut from the same gold material and glued in the top section, with a silver button for a center. It is unusually eye-catching and attractive.

Bookmarks

Everyone knows that dog-earing a book is a "no no." But how do you mark that page when you're interrupted and must stop reading? People who read a great deal usually have more than one book "going" at the same time. So why not make lots of handy little markers to mark stopping places?

I did not originate the pattern for this handy little bookmark; it has been around a long time. Many years ago we were taught to cut off the corner of an envelope to use for a bookmark. I have

seen markers made from felt, which lends itself to perky decorations. However, for everyday, constant use this new version made from contact paper is very useful and wearable.

The pattern is merely a half circle, the straight edge folded over to meet exactly in the middle, making a square corner. The shape is then a quarter circle. Do not remove the backing from the contact paper.

Cut a small strip of contact paper about ½″ or ⅜″ wide, slightly longer than the length of the open edges to be sealed together. Remove the backing from the strip of contact paper and fasten the folded-over edges together. Neatly trim off the square corner of the strip just affixed. With pinking shears (if you have them), trim the quarter circle of both sides folded together, including the protruding strip just affixed.

Another handy bookmark rescued from the trash is made from the bottom of a long plastic tomato box or basket.

Trim the bottom from the sides of the box, and that's about all there is to it.

A bow from narrow ribbon or soutache braid can be tied to a top section.

The shorter-length baskets make good markers for paperback books.

Metal Spout Butterflies

You know those metal spouts found in salt boxes and many other grocery and laundry products? I call them "butterflies."

You can make a clever mobile from these butterflies. Hang it in a child's room, the rec room, or the patio. There are fifteen butterflies in the sample, but the arrangement can be varied to accommodate a lesser or a greater number.

Bend or shape the wings so the protrusions of the notches are on the outside (underneath) of the butterfly.

Spray paint the butterflies with different colors, including black, white, copper, and gold. Some of them can be left un-

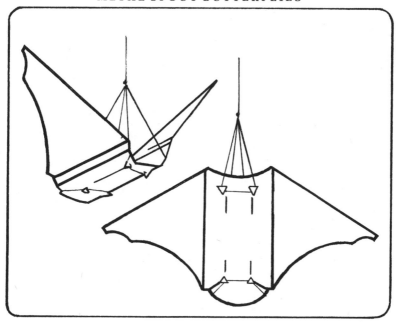

painted, if you like. I blended in Rub 'N Buff metallic colors, resulting in overall variegations and blends. You can dry brush the solid colors with silver and/or gold. If you have a steady hand, you could decorate with dots and lines, making the butterflies more realistic. I settled for the easy way out.

When you have finished painting and decorating, spray the pieces with clear plastic. Let them dry thoroughly.

For feelers, twist fine wire under the two front notches. Shape into "feeler" positions; clip off excess wire. You can use black, silver, or copper-colored wire.

Knot together approximate 4″ lengths of thread. Clip away excess ends from the small knot you have made.

Most of the metal spouts have three or four little punched-through holes and the resultant points. Hide the knot in your thread by one of these little notches; then loop the thread under the remaining notches. Bring the two loops of thread up on the outside of the metal piece and over the top of the butterfly. Pull the threads lightly to even the lengths and balance the butterfly.

For hanging, tie a length of thread at this center point of the two loops.

Some spouts have only a dent or a protuberance at the back and/or front edges. The thread can be looped around these two places in the same manner as that described above.

The top hanging piece that holds this butterfly mobile is made from three small wood dowels. You could use cut-off lengths of clothes-hanger wire. You could use a round plastic lid or some other contraption you happened to have.

For the dowel top, fasten the three sticks together with fine wire, about 2″ from the ends of the sticks, into a triangle. Fasten the wire through two sides of the X made by the sticks, then through the other two sides. Clip off excess ends.

Tie the butterflies to the top piece with 8″ to 10″ lengths of thread. One is tied at each connected corner of the sticks, others on the ends of the sticks, as well as along the space between corners.

Fasten hanging threads at each of the three corners, and then tie them to a plastic or metal hanging ring.

Glass Dome Displays

Almost everywhere one looks, figurines and whatnots and miniature arrangements are displayed under glass domes. They usually cost a pretty penny!

For a few cents and some imagination there are many ways to make display "domes" that are equally attractive.

The first thing to do is to find a suitable glass that is clear and unblemished. Then, from the assortment of lids you save, select one that fits comfortably under the glass. In some cases the lid will not be glued to the glass, but it should be practically the same circumference, so that the glass and the lid will appear to be made for each other. If the lid and the glass are not glued together, the arrangement, of course, cannot be lifted as one piece, though this is no particular disadvantage.

Paint the lid flat black or a color of your choice. Several thin coats of spray paint are more satisfactory than a thick coat that is likely to run.

Here is a description of the arrangement inside the glass with slightly sloping sides: A very small dried milkweed pod (half of it) is anchored to the lid by stuffing it with papier-mâché (prepared according to package directions). A sprig of yellow miniature straw flowers is tied together (use sewing thread or fine wire) and anchored at the stem end of the milkweed pod, where there is a partial opening. The stems of some dried weeds are gently pushed through the pod. Small portions of papier-mâché are placed on the lid next to the pod (to resemble ground or dirt), and

two miniature pheasants are glued into the papier-mâché; also, a few of the small blossoms are added.

Whatever you use in an arrangement, endeavor to camouflage or cover up whatever holds it together, so that your work will not appear amateurish.

There are certain glues and glazes that are supposed to dry invisible, but they are not always entirely satisfactory. If you want to glue the lid and the glass together, you might try one of these products. However, if the glue shows through the glass, a strip of contact paper (in a harmonizing or contrasting color) can be put around the glass, the width of the lid. Or, you could use a decorative braid or other trim.

The small slim glass: This glass is mounted over a black plastic miniature tray or dish, which is turned upside down. The glass and the bottom are not glued together. A miniature angel with a musical instrument stands glued atop a flat dried flower, which is glued to the black base. To one side of the angel is a small cluster of dried straw flowers and dried weeds.

The square jar: This is an ordinary apothecary jar, turned upside down on its own lid. It contains plastic flowers and a feather butterfly, all in orange colorings. The flowers and the butterfly are not glued.

The round arrangement, about the same size as the apothecary jar: This container formerly held candle wax. Hot water and detergent will clean a container such as this, after the candle has burned up. The jar is inverted into a red plastic 3½"-diameter lid from a coffee jar. It contains an arrangement of red plastic flowers and greenery, plus a blue glass bird in flight. The flowers and the bird are not glued.

The glass that houses the keepsake watch is inverted over a lid approximately 3" in diameter, which has been spray painted gold. A decorative wall hook for hanging pictures was taken apart and the hook glued inside the glass on the original inside bottom (now the top). The end part of the hook can be flattened and bent at right angles. If you are unable to reach inside the glass, use a long-nosed pliers to hold the hook for placement after the glue has become tacky. The decorative button portion of the hanger was glued on the outside to cover up the gluing of the hook inside.

Lidded glass container: None of the arrangement in this glass is glued. A little over an inch of white aquarium gravel is poured into the bottom. Dried weeds, plastic greenery, and small straw flowers form the background. Some petal heads (stems removed) of the straw flowers are strewn on the gravel on which the ceramic lamb stands. An arrangement such as this that is not permanently affixed can be changed from time to time.

The largest dome is approximately 6" tall and 4" in diameter, probably originally a candy container. It is inverted over a plastic lid which was spray painted flat black. Inside the dome an artificial bird sits on a real bird's nest. A spray of plastic greenery is bent and fastened across the top of the dome, with small pieces of Scotch tape. The green leaves could be glued or cemented in place.

Plastic Mold Displays

In the last chapter, Glass Dome Displays, we used various odd glasses and containers to hold whatnot and miniature arrangements.

There is another inexpensive way to make these domed displays, using small plastic bubbles designed as molds for a resin craft. The bubbles can be purchased for a few pennies at craft and hobby shops. The bases are various bottle tops, which are scrounged from your castoff supplies.

The open portion of the bubble can fit entirely into a bottle cap, or it can go only partway down, resting on the spiral protrusions.

In the photograph the two samples on the left contain dried flowers, weeds, and small strawflowers.

Push these flowers and weeds carefully through the opening; let them branch out as they will. Stems are trimmed off according to the desired placement and height inside the bubble. One bubble is glued into a white plastic cap; the other fits down into a black cap.

The third bubble contains tiny chenille balls and pearlized stamens, buds, and cloth leaves. The 1½"-tall gold cap is scantily filled with colored aquarium gravel, and the bubble is glued onto this. The stems, of course, rest on the gravel.

The fourth bubble has straw flowers in colored green sand, which runs down into the 2¼"-tall white plastic cap from a laundry product. It is glued in place. A tiny gold-painted bird is glued to the outside of the bubble.

The last bubble on the right is filled with wild-bird seed, fitted into a wooden cap from a toiletry product for men. This would make an ideal paperweight for a man's desk.

[44]

Standing Bead Mobile

Who said a mobile must hang from the ceiling? Here is one that sits on a table or shelf, has good movement in a slight breeze or when exposed to heat waves. Or, just touch it and it will dance for you!

The base is a small bottle, weighted with sand. The curved arm is clothes-hanger wire. Shape the wire into a form that suits your

fancy and that will give suitable space for hanging the beads. Anchor the wire in the bottle of sand. Use masking tape if necessary.

The bottle and wire are covered with a coating of Art Metal (obtainable at craft and hobby stores), purposely rough-textured. You could use papier-mâché, but the Art Metal adds weight to the base. It dries very quickly as you put it on.

Paint any color desired. The sample is sprayed black, dry brushed with turquoise Rub 'N Buff. Finish the entire piece with clear plastic spray and let it dry thoroughly.

The beads are turquoise, strung on invisible nylon thread. With a needle, loop the thread around the first bead and tie in a knot. Cut off excess end. Allow several inches between beads, then loop around the second bead (the thread will not show on the bead); allow a few more inches of thread, then tie to the wire arm with double knots. Cut off excess end of thread.

On the sample there are six sets of two beads each. You could have more than two beads on a string, if you wish.

Space the beads on the threads so that they hang at different intervals, with a pleasing overall effect.

Snake Charmers

Perhaps you'd rather call these creatures wigglers instead of snakes. Whatever the name, they'll charm some youngster.

Save your Leggs containers; beg more from friends. You can see that I've had quite a bit of help. When you have a goodly supply of these, you can begin to string them together; add more as they are acquired.

For the face of the creature (the sample is silver), punch a couple of small holes in the center end with a heated awl or pick. You can use either half of the egg. (I started with the smaller half.) The larger half would give a more pointed face. Bend the end of a length of wire and run it through some sort of ring; then put it through the two holes in the egg and twist the ends of the wire

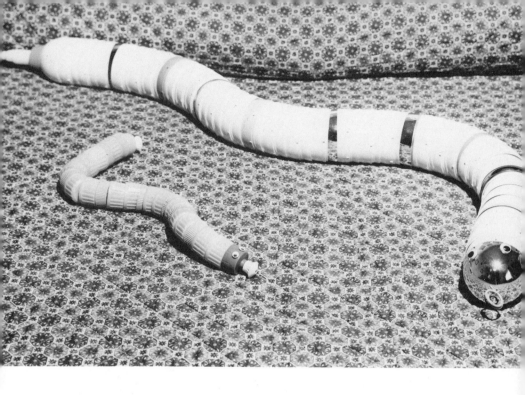

together to secure the nose or mouth, or whatever it is. The ring on the sample is a coral plastic ring used for marking chickens. A plastic or metal curtain ring would be suitable. Glue on two movable eyes (obtainable at craft stores). For a younger child you might want to use shank buttons and also affix them with short lengths of wire through other holes you punch in the proper locations. This type of eye could not be removed.

With a heated awl or pick, punch holes in the centers of all the egg halves. String them up against the head, alternating one of each kind, a larger half, a smaller half, et cetera. If you have colored eggs, intersperse them with the white ones.

When your snake is as long as you want to make it, bend the end of the wire through the last egg into an eyelet; bend and shape so the wire is secure. You can leave a slight amount of play in the wire for the entire wiggler.

Glue three or four bottle caps at the end, over the wire, simulating a tail. The sample begins with a blue Clorox cap, then has a 1½"-long white plastic and two smaller ones, the last one probably from a toothpaste tube.

To add speckles on the creature, go over the little protuberances on the large halves of the eggs with a marking pen, any color of your choice.

The smaller wiggler is made from yellow and blue bottle caps, from cleaning and laundry products. The face starts with one of those push-down caps that releases detergent, and also ends with one glued into the last body cap. The starting end of the wire is pushed through, or over, the little plastic bar on the inside of the head cap, with the ends twisted together.

The body is strung together and completed as described above.

When not being played with, these wiggly snakes would probably like to sleep on somebody's bed.

Fooling With Aluminum Foil

Sometime in the past I undoubtedly read about or was told about "antiquing" objects with aluminum foil. I did not "invent" the craft. But one day I realized what fantastic possibilities there are for glamorizing castoffs and throwaways, which of course are my stock in trade. There are many useful items to make; also, one may save a great deal of money by creating outstanding, individualized gifts.

The first thing I made, or covered with foil, was the plastic poodle dog shown in the photograph. I later discovered that a shape such as this is a bit more difficult to work with than flat surfaces, such as those of a box. Because of the rounded, shaped portions there is more piecing to do, but the dog turned out satisfactorily anyway. It was originally a bank, which I inherited after it had become dilapidated enough to be thrown away.

Although the poodle is an excellent doorstop (weighted with sand inside), he trotted off to my daughter's home, to join scores of other dogs in her collection. The dog had a hole in the bottom, besides the slot on the back of his neck. These holes were closed with masking tape after the sand was poured in.

Following are general instructions which apply to any article with which you are working: Use as large a piece of foil as possible. Be careful not to rip or tear it. If you are covering a box, cut the edges

straight and allow ample material to go around, or over, an area. You can trim away any excess. If you have to piece the foil (as with the dog), it is better that the edges are not cut so precisely or straight, because then they can blend or meld together better with another piece which they overlap.

If you are covering a box, the inside should be painted first and allowed to dry. Or, it can be lined with felt, velvet, or other suitable material. If you paint it black (the same as the outside), all the painting can be done at the same time. If the box is lined and there are areas that would be ruined by paint spray, the black paint on the outside will have to be brushed on instead of sprayed.

Decorations can be cut from cardboard. Petal flowers can be glued on in layers, but do not let the motif become too thick. Other objects can be used for decorating, but they should not be much thicker than ⅛″. If a decoration is too thick, there is a tendency for the foil to be poked and torn. Heavy string or cord make curlicues. All decorations are glued in place and allowed to dry.

Crumple the foil slightly. It is well to use a brush for applying glue. Or, you can use a spray adhesive on some areas or objects. Cover a section at a time that can be handled easily. Carefully press down any indentations around the raised decorations. Do not spread the foil entirely flat—you want some of the crinkling to remain. The foil adapts easily to round edges, odd spaces, et cetera, as it can be pushed into place to cover an area. When you piece and overlap (only when necessary), be certain all edges are glued down securely and show as little as possible. Work with a damp cloth to keep fingers and foil wiped clean.

When the item is covered with foil, let it dry first, if you can keep yourself away from the interesting project. I have worked very satisfactorily by continuing right on with the painting. Brush or spray paint with flat black enamel. When the paint is just about dry, use a clean, soft cloth to wipe and remove the surface paint carefully. You can also use paper towels for wiping. A little experimenting will show how much paint you want to wipe away—that is, the proportion of silver and black needed to give a desired antique look. You can repaint and rewipe areas if the paint has not completely dried.

Finish all surfaces with two or three light coats of clear plastic or varnish spray.

To go back to the dog: After it was finished, a circle of felt was glued on the bottom. It looked okay without eyes, but sometime later I came across a couple of silver buttonlike gadgets, which were affixed for eyes. I felt a little better about his being able to see!

The can could be used for a pencil holder, or for any items of your choice. It is painted black inside, with a circle of black felt glued in the inside bottom. The outside bottom was finished with the foil, the same as the sides of the can. The circle of foil pushed down nicely into the bottom rim of the can. The decoration is a portion of a plastic tomato basket cut out and glued on. Spring clothespins are excellent to hold something like this in place while the glue is drying.

Shown in the photograph are two halves of a box—one finished, one in process. Actually, these are two small wooden trays or boxes of the same size. I think they were used originally as dividers to hold merchandise in a variety store. They will require hinges and a closure or lock. These fixtures were salvaged from a discarded box. If you are finishing a box that already had the metal fixtures, these should be removed and replaced later. You will note the decorations on the sides of the boxes are round toothpicks and plastic rings. The decoration on top was made with metal flowers, petals pounded flat. They were an old piece of jewelry, torn apart. The centers are two sizes of small plastic rings. The curlicue is twine or string.

The large jewelry box is one I had for years. I had covered the entire top (which has carved flowers) with jewels and beads and braid. This decoration had become so dilapidated the box begged to be thrown away. But I wanted to renew it in some manner. I soaked off and dug away all the glued-on stuff, getting down to original wood. The top is actually not very adaptable to the foil covering, as the flowers are too thick and the shapes do not show well. However, the box was "newed" up and didn't turn out too badly. The sides adapted better to the foil covering, as the raised portions are not so high.

The inside of the lid is covered with one piece of foil. The inside of the box is lined with velvet. The lid overhangs in the front for opening. The lid was attached to the box with a continuous hinge across the back. This, of course, was removed while the box was being refinished.

It is extremely satisfying to discover the many throwaways and castoffs that can be reclaimed and beautified with this application of aluminum foil.

Go ahead and "foil" your friends!

Giant Paper Clip

Isn't it frustrating to look for an important letter or bill that needs attention, only to discover you can't lay your hands on it?

Here is a gadget to help clear up the clutter on your desk and perhaps save embarrassment because you forgot an important date.

The giant wire clip is shaped from a clothes hanger, then fitted into a weighted base. It is well to use bronze- or gold-colored wire hangers, as the black enameled hangers are likely to chip.

The sample wire clip is approximately 7" tall, plus the length necessary to anchor it inside the jar. You can make it shorter, or any length to suit your personal needs.

Looking at an ordinary paper clip will make it easier to visualize the proper shape. Use something substantial, as a hammer handle, around which to bend the wire. Though pliers are an aid, do not use them directly on the wire, for they will leave marks. Put a couple thicknesses of paper toweling over the wire when you are using pliers.

First, clip off the hanging hook and the double twisted wires on the hanger. Then straighten out one curve of the hanger.

Looking at the diagram, begin at A, the smaller inside curve. This can be one of the bends or curves already a part of the hanger. Cut away the excess wire, leaving the shortest loose end at B. Then continue on around from A to shape the entire clip. Bend and twist the wire so that it lies flat and straight, trying to keep it this way in the first place. The vertical lengths of wires should be spaced apart evenly.

Bring the wire around to the bottom curve, with a slight bend

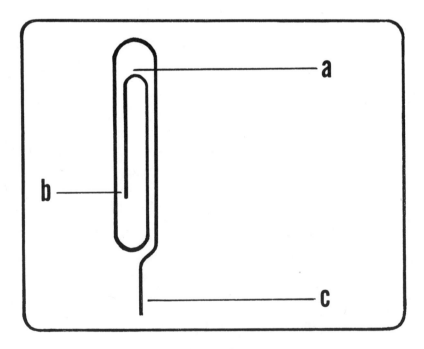

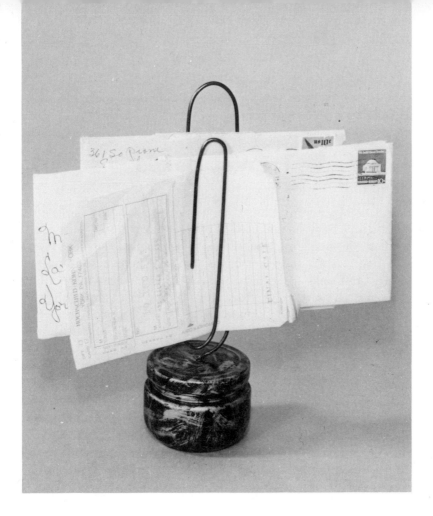

before the final straight piece that fits inside the weighted base at C. Leave as much wire at the end as is necessary to fit inside the particular jar you use. The jar in the sample held pimentos. It is approximately 2⅝″ in diameter.

With a nail punch a hole in the center of the lid, slightly smaller than the coat hanger wire. Fill the jar with plaster of Paris, mixed according to package directions.

Screw on the lid, insert the C end of the wire through the hole. Hold the clip in place, balanced straight, as the plaster of Paris gets hard.

With a piece of paper for protection over the wire, spray paint the jar. Two or three thin coats are more satisfactory than one thick coat, which is inclined to run. The sample is painted flat black.

When the paint is thoroughly dry, dry brush with gold or any of the Rub 'N Buff metallic colors. Spray with clear plastic or lacquer for a protective coating.

Trinket Boxes

Scarcely anything hurts me more nowadays than to see the plastic caps from the many kinds of spray cans thrown away. They are made in such attractive colors and can be put to so many uses.

Here are some trinket boxes that will hold any number of things: cuff links, rings, small jewelry, buttons, sewing notions, thimbles, coins, stamps, desk supplies, such as paper clips, et cetera.

The trick is to use as the bottom of the box one of the caps that has a protruding or an overhanging rim. Another straight-sided cap of approximately the same size fits as a lid onto the rim inside the bottom cap.

Decorate the boxes with odds and ends, to suit a particular person or a specific use. Or, you can use them plain, without any decoration. Men will like them not too fancy.

In the photograph: Purple bottom is decorated with darker purple braid, top and bottom. The silver domed top is the larger portion of a Leggs hosiery container.

Light-blue bottom has fancy silver braid glued around the rim. The white top has a white and silver earring (backing discarded) glued in the center.

Light-purple bottom has gold braid glued around the rim. The white top is decorated with a gold self-sticking seal or medallion.

Light-blue bottom has gold self-sticking craft braid attached. The white top has a flat piece of discarded jewelry with a blue stone in the center.

Light-pink bottom has blue/green braid glued around the rim. Darker-pink top has blue pushpin (from the bulletin board) pushed through the center for a handle.

Black bottom has black fringe glued around the rim. Black top is decorated with silver self-sticking seal or medallion.

White bottom has self-sticking craft braid attached. Orange top has an earring of white stones (back discarded) glued on.

Pink bottom, plain black top with small ceramic or plaster duck glued on for decorative handle.

One of these boxes could be the container to hold a small personal gift, such as hankies or a silk scarf, to take to a hospital patient, or as a gift for a special occasion. Hard candies in one of the larger boxes would be a nice surprise!

Uncluttering Kitchen Clutter

Undoubtedly there are unused items rattling around in your kitchen cupboards and drawers. Let's see what we can do with some of these clutter-uppers.

Meat Grinder Vase: Underneath the bouquet of flowers, do you recognize this as a table-model meat grinder?

A friend who borrowed this grinder had broken the blade most often used. I was unable to replace it. The crippled grinder cluttered up a cupboard for months, and when I purchased an electric meat grinder, I set out to give the old one some purpose in life besides getting in my way. Here it is used as a vase or flower container.

The handle with the spiral pusher was put aside for the time being. It might turn up later in another project.

The bowl of the grinder was turned around, to leave more space on the opposite end of the base. One cutting blade was attached to the grinder, in its usual place, with the regular screw-on ring (this is in back). The hole in front (that accommodated the handle) was covered with an ornamental silver button, glued in place after the metal bowl had been sprayed with silver paint. Another color button could be used, glued on before painting.

On the base of the grinder the brand name appeared in raised lettering (where the girl stands). This lettering was filed down with an ordinary flat steel file. The base was originally white enamel. I repainted it yellow. The entire grinder could be painted one color, if you prefer.

The ceramic girl with the umbrella had been given to me by my daughter with some of her discards. It happened to be yellow—just the color needed for this project. As a perfect touch, the girl's eyes are looking upward, so she appears to be gazing at the flowers overhead. The plastic flowers are mainly yellow and orange, some white. Don't be skimpy with flowers.

The figure, of course, adds interest, but is not essential. You might have, or would want to get, another appropriate decoration, such as a ceramic cat or dog or other animal. This can be glued in place with any reliable glue.

Flowering Knife Holder: The container holding the bouquet on the wall was originally a slotted knife rack. It was painted white, had a decal on the front. A decal can be removed by soaking or by sanding it off. I repainted the holder the same white color, after removing the decal. The slots are perfect for holding artificial flowers.

The flowers in this bouquet are made from *soutache* braid (on a daisy wheel) and from raveled yarns (pompons). (Several spools of braid had been picked up at a rummage sale for a few pennies, just to be on hand for *some* later project.) Raveled yarns (wool or rayon) make far more interesting pompons than new yarn, for they give an agreeable textured appearance. (Making these flowers is another project.) Other flowers that could be used in the knife container are: plastic, dried, wool yarn, straw, et cetera. Add a few sprigs of greenery, something feathery or fluttery, in keeping with the type of flowers used.

Iced Tea Jewelry Rack: After all the drinking glasses have been broken, what do you do with that handy storage/carrying rack? Here is one, turned into a useful, decorative rack or holder for costume jewelry—pins, rings, bracelets, necklaces, cuff links, scarf holders, watch.

This particular rack held iced tea glasses, so it is taller than some holders, but the idea can be adapted to whatever sort you happen to have.

Select areas of the rack that will hold a padded tray. The sample shows three—an oblong on the top and a half circle at each end. One of these trays has been removed from the rack so that it can be seen more clearly. Other trays could have been added, such as one through the middle of the rack.

From sturdy cardboard cut a shape (outline it first on scrap paper) to fit the section to be covered. Pad the cardboard lightly with cotton and cover with velvet or other suitable material. Fasten the material with needle and heavy-duty thread, crisscrossing stitches back and forth on the underside. To finish the underside of the tray neatly, glue on a piece of felt. Decorate the tray with gold braid or other trimming. This gives a rim effect.

Bird in a Whisk Cage: I was going to be a gourmet cook and just

HAD to have a whisk beater for some glamorous dish. For some reason the beater was never used. Here it is used as a decorative bird cage.

Make a perch from a piece of pliable wire, bending the ends around opposite sides of the whisk, around one of the wire loops. If the bird doesn't want to remain in an upright position when glued on by its feet (it must balance), cut a small slot in its "tummy" with a razor blade. Put glue in the slot and straddle the bird over the wire perch.

The posies are clothlike flowers from an old hat. You could use small plastic flowers. Wind the strands or stems around the handle and fasten as you go by wrapping or winding with fine wire. A couple of dropped blossoms are glued on below the handle. A whimsical butterfly is attached by its own wire atop the flowers.

Incidentally, if you wanted to buy a new whisk and decorate it, this would make an attractive gift for a shower, especially a kitchen shower. Whisks come in various sizes and the size of the bird used can be adapted.

Hamburger Press Plaques: Why I bought a second hamburger press I'll never know, for I had previously discarded one never used. This time, when weeding out the kitchen, I decided to make a couple of hanging plaques from the unused press.

After taking the two pieces apart, the larger piece was stained. If the wood's finish is in good condition, without decals or decoration, you might not have to do anything to refinish it. After the stain was applied, I rubbed on a bit of furniture wax. The piece of metal which is part of the handle remains the original black.

A small mound of florist's clay was put in the center of the flat side (the shaping depression is on the back side): dried flowers and weeds were arranged in the clay. For a more permanent arrangement you can use papier-mâché or a clay that hardens. In this arrangement I used acorns, dried goldenrod, strawflowers, thistles, a sweet-gum ball, and a couple of other weeds.

The hanging hole at the top was covered with an upholstery tack. Put a little glue-soaked cotton on the underside of the tack and glue it in place. A portion of the hole remains at the back for hanging.

The smaller part of the press was spray painted light blue, including the piece of black metal in the handle. A flat piece of discarded jewelry from a bracelet or belt was glued over the hole in the handle. This happened to have a blue set in the center which matched the paint.

Instead of using the flat side of the press, the raised circle (that fit the depression in the other piece) is the front. A photograph (my daughter in her teens) was cut to fit the circle and glued in place. Then gold braid was glued around the outer rim of the photograph.

A small slice out of the wood at the top accommodated one side of the hinge that connected the two pieces of the press. There are three screw holes here. A piece of gold braid for hanging was tied with a knot at the top and then glued across this flat indentation.

Omelet Pan Plaque: The double or "twin" wall ornament probably gives itself away promptly as a former omelet pan.

Nothing was done to the black handles (which are in good condition), except to remove them in order to paint the long screws holding them in place. The screws and the balance of the pan were sprayed with silver. Any color of your choice could be used, to match or carry out a color scheme.

Decorative arrangements in the pan can be made more or less permanent by using papier-mâché or molding clay that hardens, to anchor the flowers. If you use florist's clay (as I did), the arrangements can be easily changed, perhaps to feature the four seasons or for certain specific holidays.

Two different kinds of arrangements are shown as samples, but your arrangements should be more matched or co-ordinated.

One section of the pan contains dried weeds, pods, and nuts, with a glittered bird perched on the gold-sprayed unshelled walnut. In the second section of the pan are two sprigs of flocked pears and a large flat plastic flower.

The holder hangs on two nails or two decorative hooks.

Stationery With Iron-On Silhouettes

Almost everyone loves pretty stationery. A special love note sounds even sweeter when written on one of your own originals.

There are countless ways to decorate stationery. Some crafters specialize in this hobby alone. Often the work is painstaking and tedious. A long time ago I discovered a way that is easy and uncomplicated to perk up a piece of paper. Simply cut silhouettes from ordinary iron-on mending tape. This tape comes in assorted colors, in rolls or in larger flat pieces.

What is especially interesting is that you can use motifs to fit almost any person or any occasion. Cutouts can be suggestive of hobbies or interests, or they can be merely decorative. A pair of small sharp scissors is the only tool required. Cuticle scissors work well; they go around curves easily.

The size of the silhouette cut-out should be adapted in size to the paper or card on which it is to be used. Besides using the cutouts for stationery (which can be a single sheet or the double-page type), you can decorate recipe cards, gift cards, name place cards. Paper or cards can be white or tinted.

Blank visiting cards (2⅛″ x 3½″) can be purchased at variety, department, and stationery stores, either with or without envelopes. If you want a smaller card, as for a name place card, you can cut down the size. There are also large-sized correspondence cards available (3¼″ x 5⅛″) if your message is more lengthy.

Do not worry if the silhouettes or outlines are not perfect; you do not want them to appear stilted or machine-cut. With a little practice you will find it simple to look at the outline of an object and cut as you go. If you feel a bit uneasy about doing this, experiment by cutting samples from paper before using the tape. If necessary, an outline can be traced on the wrong side of the tape. When drawing an outline on the back of the tape, remember that it will be reversed when applied.

Preferably use not more than two or three colors for each design. The colors, of course, should blend or contrast nicely. It is not necessary to follow the natural color of an object, but this can be done as nearly as possible, if you wish. With some silhouettes the natural colors give authority to the shapes of the objects. Always keep the outline simple. For a bit more realism a few lines can be drawn with a pen, as on the fruit and the bottles, but this is not actually necessary.

To apply the cutouts to paper, use a warm iron (no steam) to iron them in place on the card or stationery. Hold the iron steady while the cutouts are being applied. Do not move it back and forth vigorously. If you can't affix all the pieces of one pattern set at the same time, iron on the pieces separately so they will not become disarranged. When the iron does not cover the entire surface of the paper, be certain to hold the iron lightly, moving over the entire surface of the paper. This is to avoid iron marks.

If you want to individualize a personal gift, it might be interesting to make a gift card with the silhouette matching the gift, such as a tie, a bottle of perfume, a plant. Friends' personal hobbies supply endless ideas: fishing, boating, bowling, card playing, painting, et cetera. Holidays, anniversaries, showers—the list can go on and on.

Besides decorating stationery and cards for your own use, you could make some as gifts for friends. A dozen or two gift cards with envelopes, for instance, can be packaged in a small box (cardboard or plastic), or a cellophane bag. The cellophane bag can be fastened across the top with a folded piece of cardboard, stapled together near the ends. On the cardboard you might write, print, or type: "Handcrafted by [your name]."

It is a great convenience to have on hand greeting cards for unexpected occasions that are always popping up, as well as for the planned ones. A simple, sincere message written on a decorated correspondence card, or small folded stationery, makes a thoughtful remembrance more appreciated than a "store-bought" card.

You can cut a lot of capers with silhouette cutouts!

Bulbous Flower Or Sea Creature

This project requires a considerable number of burned-out flash bulbs. If you do not have a sufficient number of old flash cubes, you can begin with what you have and complete the project as you acquire the bulbs, as I did.

The "thing" we are going to make is definitely a conversation piece—bulbous flower or sea creature—who knows?

The base is a styrofoam ball. The one in the photograph is about 2¾" in diameter. You could use a smaller or a larger size. Cut the ball in half.

The bulbs are attached in the flash cubes by small wires. Wiggle them out gently with your fingers, or use a small pliers if necessary. You have, of course, removed the plastic square to which the bulbs are attached. (The plastic square is easily released by pressing gently on each of the four sides of the clear plastic.)

Some of the flash bulbs have two wires; some have only one thicker wire by which they are affixed. If there are two wires, twist them together after removing the bulb. Then stick them (as one wire) into the styrofoam ball. Just keep adding bulbs until the ball is completely filled; that is, all except the bottom.

If the bulbs are not unduly handled and their positions changed, they will remain solidly in the styrofoam ball without any fixative. If necessary, use glue, cement, or a molding clay that hardens. Molding clay is what I used as I neared the bottom, especially on the last two rows.

Cut a circle of felt the diameter of the cut styrofoam ball and glue it to the bottom.

For years I have had a couple of tiny stick-pin beetles which came from a Scandinavian gift store. The beetles are purported to be lucky. Recently I came across one remaining beetle and stuck it into the "thing." So I presume my "thing" is a BULBOUS flower!

At least, it's a blooming conversation piece!

Easels For Many Uses

Some years ago I mounted dried flower and weed arrangements on backgrounds made from gold-finished, anodized aluminum. (It is also available in silver finish.) This aluminum is sold in sheets in various sizes, in several different patterns. I like the cloverleaf pattern best. The aluminum is available at hardware and building-supply stores. I found that scraps or pieces left over from a project could be used for many other things.

One of the most useful items to make is an easel—an excellent standard or holder for propping papers and notes while you are typing. I am using one as I type this manuscript. Also, you can display a favorite photograph, a small oil painting, or a plaque.

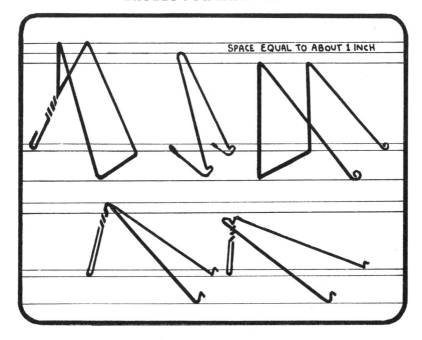

Sometimes there is no place to hang a small calendar near the desk, and an easel serves well to hold one. The standing easel can be a plaque itself to hold a flower arrangement or a special whatnot.

This aluminum is easily cut with any type of heavy-duty shears, and it shapes easily. A ruler can be used to hold and brace a piece while you are bending it.

The shapes to form easels are too numerous to detail. The samples shown will give you ideas to start with. You might draw on paper a rough sketch of how you want to cut a pattern; then count the pattern squares for size. The "feet" or bottom of the standard may be perfectly straight, or they may be bent up as arms for holders. Or, the entire front piece can be shaped without feet.

The small easel holding the photograph is ten cloverleaf patterns or squares long, bent in the middle, with one square at each corner turned up at right angles for feet. It is three patterns wide, and all but three of the patterns in the middle are cut away.

The photo is mounted in a Mason jar ring, with a circle of clear plastic or acetate for "glass." The picture is backed by the lid, taped in place, and the rim then filled with plaster of Paris. The rim is

trimmed with variegated gold braid. The photo is fastened to the easel, or picture standard, with a snipped-off length of aluminum, about ⅛″ wide, cut from a straight side of the sheet.

Other serviceable easels can be made from ordinary wire clothes hangers. I couldn't get along without several of these handy gadgets.

The drawings and photographs will give you ideas as to how the wire may be bent. There are workbench devices for bending wire and you may be fortunate enough to own one of these. I don't, so I used a pliers. A piece of heavy cloth under the pliers protects the finish of the wire. The easels could be painted various colors. One of the easels has a couple of elongated beads pushed onto the end pieces.

Patterned Aluminum Jewelry

From further scraps of anodized aluminum you can also make attractive costume jewelry. The samples in the photograph use the same cloverleaf pattern as that used for the easels. They will get you started with ideas. Possible shapes to cut are too numerous to detail.

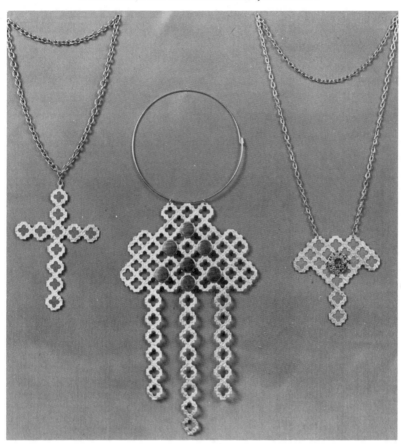

You might start with a simple cross. This is seven patterns long and five patterns wide. After you have cut the aluminum, go over the back side and edges to smooth any roughness by rubbing the hammer head over it.

The cross is backed with red felt. Cut small squares of felt to fit behind each clover leaf. Clip off the corners of a square so no felt will protrude and show from the front. Glue the squares in place.

For a hanger for the cross, cut a length of the aluminum about ⅛″ wide, 1″ long. Bend it in half, making a loop. Glue this loop to the back of the top felt piece; then glue a second square of felt over the loop. Add a couple of jewelry jump rings and a chain.

[67]

The smaller pendant-type piece also has felt glued to the back, as described above. A scrap of pink felt is cut for the entire piece of jewelry, except the two lower patterns.

With a sharp-pointed nail, hammer two small holes in the top of two of the patterns. These holes are for attaching the chain through jump rings. Be sure to center the holes and not go out of bounds of the small area where you are punching the holes.

A discarded earring with a pink stone is glued to the front of the pendant. This, of course, is not necessary.

The hanging piece on a choker has been fashioned for a large-sized person. It might overpower a petite girl. For further decoration gold buttons with black streaks (shanks removed) are glued to the aluminum.

It also has the punched holes, as described above, for attaching to the choker. The choker rings or bands are inexpensive and are obtainable at stores where craft supplies are sold.

Dangle earrings could easily be fashioned in the same manner as the neck pieces, for matching sets.

How Sweet It Is—A Lemon!

Don't throw away those empty plastic lemons after the juice is used. They can be turned into interesting and useful "personalities."

We can make a pencil holder, a paperweight, a flower holder or planter, and a piggy bank (or just a piggy).

Flower holder or planter: With a sharp knife cut off about 1″ of a lemon at the end opposite the opening. Leave the cap on the opening.

Mix a little of plaster of Paris according to package directions and pour into a plastic cap approximately 2″ in diameter. Do not fill the cap full. Experiment as to how far the lemon will fit down into the plaster of Paris without its spilling over the top edge. Wipe away any excess plaster with a damp cloth. Small live plants, such as cactus, can be grown in this container, if you use recommended soil. It can be filled with sand to hold plastic flowers. It can be filled

with water and used for a small bouquet of fresh flowers. The sample holds blue plastic flowers and greenery. The yellow lemon sits in a blue plastic cap base.

Paperweight: Anchor a lemon in a plastic cap in plaster of Paris, as described above. If you want further weight, some thinned plaster can be poured into the opening of the lemon; then replace the cap. Add a face to the lemon. Glue on two movable eyes (obtainable at craft stores), a flat red bead for a mouth (or a bit of red felt), and push in a black map pin for a nose. Or, use a small bead, button, or bit of felt for the nose. The sample paperweight wears a tilted white plastic cap (1¾" in diameter) to resemble a sailor's hat, glued in place.

Pencil holder: This does not necessarily require plaster of Paris in the base cap, but some can be used if you wish. A small hole approximately ¾" in diameter is cut out in the top of the lemon. For the girl or lady, the nose and mouth are the same as described above for the paperweight. The lashes (no eyes) are from a discarded pair of actual false eyelashes—one lash cut in two. Glue in place, slightly slanted. At the side of the head a fluff of blue feather and a small bow tied from a strip of blue felt are glued in place.

The boy pencil holder has a little curled-up lock of hair. Cut out the top hole for holding the pencils. Cut a sliver about 1/16" wide, tapered at the beginning, for the curl. Eyes are cut from black felt—two open oblongs and two dots glued in place as if looking up at the bee (available at craft stores) poked into the side of his head. His mouth is cut from red felt. The tie is a piece of turquoise felt about 2½″ long and 1″ wide, pulled together in the center with a small piece of the felt. Glue the tie in place on the white plastic cap base.

Piggy bank or just a cute whatnot: If the piggy is to be a bank, cut a neat slot in the lemon with a small sharp knife. The tail is a 1¼″ length from a piece of metal pot scrubber. With a large needle punch a hole in the end of the lemon and push in the tail. You could use a piece of pipe cleaner, twisted into a curl. Or, make a small coil of wire. The four legs are yellow plastic pushpins (from a bulletin board). Small movable eyes are obtainable at craft stores. For the nose the cap of the lemon is discarded. A jewelry jump ring is fastened into the "squirt hole" with a piece of fine wire twisted around it.

I happened to have a small pink hat with blue plastic flowers, and a pin is stuck into the piggy to anchor the hat. You could make a hat from cardboard, cloth, or use a small plastic cap glued to a rim. You could crochet a hat. Or, you could add a jaunty sprig of flowers or a piece of discarded jewelry for decoration.

Several lemons can be co-ordinated into an attractive set for a special person's desk—perhaps your own.

Imitation Shell

Almost everyone who examines one of these "pictures," as shown in the photographs, thinks the flowers are made from shells. Not so. These are lowly plastic spoons and forks, heated and softened over a candle flame.

One of the most suitable and attractive backgrounds is velvet. Other materials can also be used, such as upholstery fabric, as shown in one of the photographs. Strips of the same background fabric were cut and glued to the frame. The background in the other photograph is red velvet. The purchased frame already had red velvet on the frame—a perfect match.

You will need an artist's canvas board, a piece of lightweight plywood or masonite for the backing.

The fabric should be cut 2″ or 3″ larger than the backing all around. It can be fastened to the very edges of the backing with staples (be certain the staples do not show at the inside edges), or it can be securely glued at the back edges. But it is also advisable first to "sew" the fabric together on the back side by crisscrossing back and forth with needle and double thread from side to side and end to end. There is considerable weight on the cloth fabric when the flowers are mounted, and securing the background in this manner will prevent sagging.

To affix the flowers and vase to the cloth background, I prefer using dabs of art metal (obtainable at stores where craft supplies are sold) or a molding clay that hardens. If you use glue, it should be an epoxy type, as this has greater holding power than some other adhesives.

The flower arrangement can be all one color (those on the red velvet are white only); or, you can use a mixture of colors (as on the green upholstery). If you use a mixture of colors, green spoons and forks, of course, are the natural choice for leaves.

It is well to cover your working surface of newspapers with foil, so the plastic will not adhere when it is warm and soft.

Experiment holding a spoon over a candle flame. You want the plastic to be smoked, but you do not want it to be charred or burned. If too much soot accumulates, sometimes the excess can be wiped away. Cut off the softened spoon bowl from the handle and quickly do any shaping necessary before the plastic hardens. The bowls will usually soften into natural shapes. Use the points of your scissors or a small tool if necessary to manipulate the pieces into indentations et cetera. Once a piece hardens, it cannot be shaped or reshaped. You will have to soften it again. Make up a number of the single petals to form later into flowers.

When you hold the tines of a fork over the candle, they will curl into various shapes. Cut the tines from the handle. Usually you do

not have to do anything further. But if a tine gets too far out of place, or curls in a manner you do not want, soften again and push it into place. Hold it a second or two with scissors or a small tool until it begins to harden.

The ends of handles are also used, either flat or curled up. Small pieces or chunks of the plastic are melted for fill-ins.

After you have a supply of anticipated requirements, begin to assemble petals into flowers. A center can be the same color as the petals or a different color. Heat a small chunk of scrap plastic, form it into a circle, push the petals into its sides, or place the softened plastic on top of the petals. Shape as a flower. In the same manner, fashion fork tines, softening and pushing the base ends together.

The photographs will give ideas for shaping flowers and placing them as a bouquet. Experiment with all types of shape combinations. Anything goes! For the vase put a bunch of the scrap pieces (all colors) close together in an approximate shape of a container. Place in a heated oven (approximately 375°) on a flat surface covered with foil. When they have softened, the pieces will run together and you will have a variegated vase. Shape a little, if necessary. Another container can be made by placing two plastic spray caps close together, melt in the oven, then shape while still warm with scissors or a small tool. Be careful not to touch the hot plastic with your fingers.

Practice placement of finished flowers, leaves, et cetera, on the background you have prepared. When you are satisfied with the arrangement, affix it in place.

Frame your masterpiece and expect to hear compliments!

Jeweled Pigtails

How about making a "cutie pie" to hold that assortment of costume jewelry that is usually cluttering up a drawer and getting tangled?

The base of the face is paper plates. According to the sturdiness of the plates you are using, put two or three together to give substantial strength. The plates in the photograph are 7″ in diameter.

Cut a circle of cloth a couple of inches larger all around than the circumference of the plates. The sample is blue velvet. Almost any closely woven material is satisfactory. Make a running stitch with double thread around the edge of the cloth. Put the bottom of the plates against the wrong side of the material. Draw up the thread and fasten by weaving the thread back and forth across the cloth edges. Securely backstitch the thread to fasten it.

The face features are cut from felt—red for the nose and mouth, black for the eyes. One eye is winking. You could use discarded false eyelashes.

The hair is braided yarn. The sample is red rug yarn, because that is what I happened to have on hand. Wool is equally suitable but requires more strands. Cut lengths of yarn about 1¼ yards long. This gives you ample to work with to assure braids that will be long enough. With the rug yarn I used eighteen lengths—six in each braid section. With smaller wool yarn you would need more. Keep the three sections separate, tie the ends together, and use a safety pin or large needle through the knot to anchor to an upholstered chair, the bed, or something similar. Braid neatly. Hold the braid against the face to see exactly how long you want the pigtails. The pigtails are tied near the ends with yarn or ribbon. Trim the ends of the yarn neatly. The trimmings can be used for bangs.

Put glue on a small section of the top of the head and affix a few bangs. Then glue on the braid, starting at the top middle, working down the sides to about the center edges of the plates.

The hooks at the bottom edge of the face are fancy ones for wall hangings—they have a decorative circle button that snaps on and off to hide the nail hole. These hooks were fastened to the face with fine wire through the nail hole, tied in back. You can use large kitchen cup hooks if they have the "collar" above the screw end. Cup hooks actually hold more necklaces than these fancy gadgets. If cup hooks will not hold of themselves, use glue. Or, you can fasten them in the back with wire.

Cut off another paper plate about ½" or ¾" around the entire edge, to make it smaller than the face. Glue this plate to the back to finish it neatly.

With a double length of yarn in a large darning needle, make a stitch about 1" long in the braid about ½" from the middle top edge. Place the stitch in the braid so that it will not show. With both ends of the yarn in back, leave a small loop and tie it securely. This is for hanging the plates.

Stick jewelry pins in the top hair. Earrings go down the pigtails. Necklaces or chains hang on the hooks.

Even if you don't use the "cutie pie" for holding jewelry, she makes a clever decoration for a girl's room.

Tin Lid Pendants

What would we do without tin cans? After they hold all the things they do, we can still use them in many ways—the lids too.

You can make an attractive pendant by dressing up an ordinary tin-can lid with paper clasps and jewelry jump rings.

If you don't find a few paper clasps lying around in a forgotten place, a box is inexpensive to buy. You might like to use a larger size than the ones shown, or perhaps a combination of two sizes. I used the ones I happened to have on hand.

Select a bronze can lid that is not marred or scratched; without indented printing. Price markings can be removed with lemon extract. The sample pendant is about 2¼″ in diameter, but the original lid was larger. You want a flat edge as an outside rim, without the ridged rings that are sometimes in a can lid.

The rims can be used as a guide for cutting away the excess. Use a sharp shears and very carefully, very evenly, leaving no jagged edges, cut out a neat circle.

You need a belt eyelet puncher to make holes in the lid. The holes should be spaced evenly. Cut a piece of paper the same size as the lid. Fold it carefully in half and place it on the lid. With a pencil make light marks to indicate where the straight diameter line would be. Make the marks about ⅛" from the edge of the rim to where the center of the hole will be. Punch two holes on this imaginary straight diameter line. You might want to practice first on another lid, or at least on a circle of paper.

From center to center, the five holes are each about ½" apart in a circle of paper to use for a pattern or guide. Center one hole over the first hole you have punched in the lid. Mark the next hole from the circle of paper pattern. Do the same with the remainder of the holes to be punched.

The center dangle of the pendant uses four paper clasps and three jump rings. The next uses three paper clasps and two jump rings. The shortest dangle has two paper clasps and one jump ring.

The jump rings in the sample happen to be silver, since that was the kind I had on hand. They make an attractive contrast with the bronze lid and clasps. If you purchase jump rings, you might want to buy gold.

Open out the prongs of the paper clasps. The longer pointed end of the paper clasp is left dangling at the bottom end of the dangle. With a needle nose pliers, bend back the other prong and securely clasp it around a jump ring. Note the distance on the prong from the edge of the center button where you place your pliers, so all the prongs will be turned under evenly. Make the five sets of dangles.

The last prong of the dangle should be clipped off a bit before fastening it to the lid, so that it will not protrude and show beyond the lid. Put the clipped-off prong through a hole in the lid and turn it under securely. The edge of the center button should be at the edge of the lid.

The top hole for holding the chain uses one paper clasp and two jump rings.

You might make your dangles longer than the sample, or you could devise other patterns for assembling. Too much decoration of paper clasps can tend to be overpowering.

Earrings are easily fashioned in the same manner as the pendant. Make the dangles as long as you wish; then fasten with a jump ring to the type of earring you wear—pierced or clasp. Jewelry findings are obtainable at craft and hobby stores.

Butterfly pendant: This is a tin lid approximately 3″ in diameter, cut out neatly with an electric can opener.

With a small hammer pound the lid slightly on the wrong side, making noticeable indentations. The pounding will also shape the lid a little to convex-concave.

Punch a hole (an eyelet punch will do the job) near the edge to hold the jewelry jump ring, through which the neck chain will pass.

The decoration is a "butterfly"—the metal spout from a salt box or other kitchen product. The original wires become the feelers; or, if this wire is not satisfactory, or if there is none, you can make feelers by twisting fine wire into place.

Glue the butterfly to the center of the lid. Spray paint black (or a color of your choice) on the top side only.

While the paint is still wet, use a paper towel (a couple of thicknesses over a finger) to wipe away patches of paint, letting the silver show through. This gives an antique look. To finish, spray paint with clear plastic.

Tinsel Bouquet

As a general rule we think strings of tinsel belong to the Christmas holiday season. This need not necessarily be so. Give the tinsel another chance.

To me one of the most gorgeous tinsel color combinations is the bright shining green and blue. I made some flowers with this tinsel, combining them with small blue tree balls. The result is startling. This bouquet could be used at any time of the year. Also, gold and silver, for instance, would make fabulous bouquets for silver and golden wedding anniversaries. Other colors would carry out other party themes.

Take a strand of tinsel approximately 12″ long, more or less, gather it up into three equal loops, and tie together with matching

thread. Use an adequate bunch (not skimpy) of pearlized stamens (obtainable at craft stores) for the center. I had some light blue on hand, which made a perfect combination.

Bend the end of a length of floral wire into a U. Loop the U over the middle of the flower, with the stamen center in place, and twist the ends of the wire together. Slide on a green plastic calyx and glue in place.

If you have used plain wire instead of covered wire, wrap the stems with floral tape.

To anchor a ball onto a stem, make a small U at the end of a length of wire, bend it horizontally, attach a wad of glue-soaked cotton, and insert it into the ball. Press firmly into place and let it dry. Add a plastic calyx to the ball also; glue in place.

The flowers in the photograph are arranged in a pansy bowl (small bowl with slots). I wish you could see them in color.

Old Belt Sewing Kit

Who hasn't juggled from here to there a drawerful of discarded belts? You know you'll probably never use them, but how difficult it is to throw them away!

Here's an idea to use at least one. After you've made your own sewing kit, you might want to make several for friends.

This handy little sewing kit should hang by your sewing machine, if you have one. If you don't have a machine, put it in

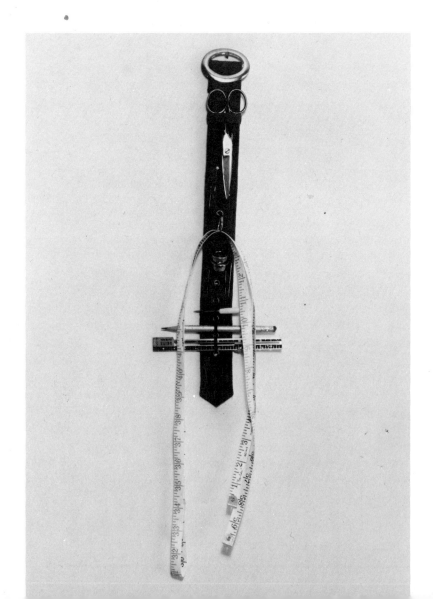

another convenient place—perhaps on a closet door. There's always a rip to mend, a hem to fix, or a button to be sewed on.

A sturdy cloth belt about 1½" or wider is preferable. You can use plastic or leather, but it will be necessary to make a separate little gadget to hold pins and needles. With the cloth belt, pins and needles are stuck right through the cloth.

Remove the buckle from the belt and cut off the portion you will not use, leaving about 17" from the eyelet end. This length will accommodate a small pair of scissors, which you are going to hang in the belt loop. If your scissors are quite large, you might want more length than 17".

With pliers remove the buckle prong, then replace the belt through the buckle, turn down about 1¼", and tack together neatly in the back.

If you have a loop that held the end of the belt, fine. If not, it is simple to make one. Clip out a piece of cloth from the discarded part of the belt. Fold a piece, both rough edges under, and stitch it together. The finished loop should be about ⅜" wide. Fasten it to the back of the belt about 2½" from the top, whipping along both sides of the loop. This loop is to hold the scissors.

In the sample, café curtain rings are used to hold the tape measure and the thimble. These have the extra little circlet with which to fasten them. A plain metal or plastic ring will do as well. Whatever ring you use, it should be of a size to keep the thimble from slipping through. The café curtain ring was a bit too large, so the end of the big circle was merely pushed through the small circle a little, to diminish the circumference.

The ring for the tape measure is whipped to the belt so as to stand vertically. The ring for the thimble can be whipped on either way, vertically or horizontally.

You will not be able to stick pins and needles in a plastic or leather belt, but you can make a separate little pincushion. If you have enough belt material, use it. If not, select a piece of soft, sturdy cloth. Cut it the width of the belt and about 4" long. Use pinking shears if you have them. Fold the piece of cloth lengthwise. At the folded side attach it to the belt with one or two small buttons, sewing through the buttons, doubled thickness of the cloth, and the belt.

If you have eyelets in the belt you are using, so much the better.

Take lengths of narrow elastic (approximately ⅛″ wide), put each end of the elastic through an eyelet hole (two adjoining eyelets), and tie securely in back. Sewing gadgets, such as a ripper and a gauge, can be slipped through these loops.

If you do not have eyelets in your belt, you could tack down loops on one continuous piece of elastic stitched to the belt.

If the hanging belt has a tendency to stick out from the wall at the bottom tip, a small piece of masking or double stick tape will hold it solidly to the wall.

Stamp Dispenser

We have all seen people buying one or two stamps. They seem to be always OUT of stamps and having to make an unnecessary trip to get one of these very necessary little items.

It is more of a financial investment at the time to buy stamps by the roll, but I have been doing this for years and wouldn't buy them any other way. You'll like the secure feeling of knowing you have a postage stamp when it is needed.

You can make a handy dispenser in a matter of minutes. It looks attractive on your desk when the color matches or harmonizes with other desk accessories.

All you need is a plastic cap and a metal lid. The black plastic cap in the photograph is 1⅞″ in diameter. The trick is to find a lid into which this cap will fit perfectly. I used a Vlasic pickle jar lid, which has four protrusions on the inside rim of the lid. It is approximately 2⅛″ diameter on top. You'll probably have no difficulty in finding a lid from still other products, but it should have the protrusions on the inside rim to secure the cap.

If the lid is not scratched and is an appropriate color (with no advertising showing), it can be used as is. I painted an old lid flat black and dry brushed it with gold (when the enamel was thoroughly dry). If you like, you can spray with clear plastic as a final protective coating. Cut a circle of felt to glue on the bottom (formerly the top) of the lid.

The slit in the plastic cap from which the stamps are dispensed is made with a heated knife or similar tool. I have an old all-purpose blunt knife that is slightly thicker than most ordinary paring knives. But a thinner knife can be moved slightly to widen the slot.

To heat the knife, I laid the blade directly on the electric stove unit. Start at the bottom of the cap and gently work the heated knife toward the top. You might want to practice first on another plastic cap. Use a stamp to check the correct length of the slot. The slot will give slightly more length by not pushing the cap entirely into the lid. Actually, it is held in place perfectly by the lid protrusions.

Now go out and buy yourself a roll of stamps and stop worrying about never having a stamp on hand when one is desperately needed.

Or, think what an appreciated gift this dispenser would make, especially when it is filled with a roll of stamps. It is ideal for anyone who *doesn't* have everything!

Strike Up A Match!

Anyone who likes to burn candles and incense as much as I do needs a handy supply of matches.

Here are a couple of ideas to make an interesting match holder and match striker.

The 4″ x 3″ stained wooden base in the photograph at one time held a bowling trophy. The front edge happens to be slanted, but

this is not particularly necessary. A piece of fine sandpaper is glued here for striking matches. Strips of sandpaper could be glued on all sides of the base. Another match striker is a small rock. You might not want both strikers for the same holder, but they are shown to give you the idea. The rock striker is glued to the base, as well as the silver cap (from a cosmetic product) to hold matches.

Going a step further, you might put your match holder and match striker on the lid of a small box which could hold incense, for example.

The top of the plastic box lid shown in the photograph is slightly indented. A piece of blue felt is glued in the indentation, with blue braid around the edge of the felt. The match holder is a straight-sided white plastic bottle cap 2″ high, glued in place. The striking rock is also attached with glue. A blue glass bird in flight is glued to the rock.

Instead of an entire box, a suitable lid only could be used. A piece of wood can be painted a color of your choice instead of being stained, or it can be covered with contact paper. An inverted round plastic or metal lid could also be used for a base.

You might like to "dress up" an original box of matches, perhaps for a pipe smoker. Cut a piece of contact paper the width of the match box and cover all four sides, beginning and ending at the strip for striking the matches. Around the edges of the contact paper and across the edges of the striking strip, affix narrow self-sticking gold tape or tape of some other suitable color. Self-sticking gold letters on the top of the box can spell out a person's name. Letters can also be cut from contact paper.

"Empty-Headed" Pincushion

These lovely little "empty-headed" ladies usually come from florists' shops and contain fresh flower arrangements or plants. They often turn up at rummage and garage sales, as people do not know what to do with them after the flowers have wilted or the plant has not survived.

When no longer needed or used as a vase or planter, why let the lady sit around and mope? Let's give her another life to live.

She makes an excellent pincushion. A styrofoam ball is the filler for the pincushion. The size of the openings in these ladies' heads vary; so select a ball to fit the opening. With a sharp knife remove the part of the ball you will not need, cutting it according to the height of the hat you want to make.

The hat is crocheted. The only stitch used is the single crochet. As you make a continuous circle, pick up both front and back loops of the previous row, so the stitches are compact and firm. Keep the styrofoam ball handy and fit to its shape as you crochet. When the hat is large enough, if necessary, take it in slightly by crocheting a few stitches together, spaced evenly at the lower edge. Then, for the brim, make a full ruffle with chain-three loops for three or four rows.

Fit the hat over the styrofoam ball and glue to the head rim of the container. Add a velvet (or other type ribbon) band and bow or small flowers, if you like.

Most of the heads do not already have a hat brim (at least, those I have seen), as the turquoise lady in the photograph. If this is the case, crochet only the necessary crown. This one is red yarn, with a piece of frilly gold braid for trimming.

Now you have a gal you can stick pins and needles in without hexing her!

Eyeglass Case—A "Spectacle" Plus A Purse

Although I have worn glasses since I was a child, I have never used the spectacle cases given me so many times. They make good drawer clutter unless they are put to another use.

Nowadays the cloth or plastic cases are usually colorful and attractive. As a little hanging or wall vase you can brighten up a dull spot.

About an inch down from the top, on the back side, affix a belt eyelet for hanging. Or, just punch a hole with a sharp object. Tuck in some plastic flowers and greenery.

A 3½''-long glass or plastic straight-sided container (about 1'' in diameter) can be slipped inside a case actually to hold water for a sprig or two of growing ivy or a trailing vine.

Another excellent use for a discarded case is a protective holder for a pair of shears or scissors.

A snap-shut coin or inner purse can also be used for a hanging wall vase. Fasten a piece of wire to the back closing piece, or punch a hole as described above.

A purse bouquet such as this can also be displayed to advantage on an aluminum easel, such as that described on pages 65-66.

Flash-Cube Belt And Jewelry

Burned-out flash cubes are probably thrown away by the thousands, but you might be surprised at the numerous treasures that can be made from this trash item.

First, how about a link belt? Remove the brown cap or lidlike piece and take off the wires attached to it. If you want a color other than the original brown, spray paint one side and let dry thoroughly. Then paint the top side. The top side is the one that has the little raised square. Finish with a coating of clear plastic spray.

Fasten the squares together with gold jewelry jump rings, at two opposite holes. If you do not have the jump rings, it might be well to take a couple of the squares with you when making the purchase, so that you buy them large enough, at least ⅝" in diameter.

There are numerous ways to close the belt in front. You might find a suitable fastener from an old belt. You can use two brass rings used for hanging café curtains. They are about ⅞" in diameter. There are plain rings only, and also a ring with a small eyelet. Attach the rings to the belt squares with jump rings. The rings can

be tied together with a thin strip of leather (an old belt tie), or make a sphagetti strip from cloth, perhaps to match a particular costume. There are large jewelry-type safety pins (sometimes used for skirt closings) that make excellent closures. Also, try a hair barrette.

To make attractive jewelry, spray paint the squares silver, as described above. In the little raised squares glue a colored rhinestone which you may have saved from discarded jewelry. I used turquoise rhinestones, and they are very attractive with the silver squares.

One square makes an earring. If your face and neck structure permit and you like more dangle, use two squares. Attach a silver jump ring in one corner hole, then the ear fastener, either a clip or one for pierced ears.

One square on a small-link silver chain makes a necklace to complete a set.

Or, you can make an arrangement of dangles and affix it to a silver choke wire to wear around your neck. The sample shown uses eight squares. You can put them together in any number of ways. If you wish to experiment with an arrangement, do this before painting the squares, to avoid scratching the finish.

Easy Crazy Quilt

Making quilts by hand has never been my "bag." I have made several crazy-quilt tops on the sewing machine, both in strips and blocks, but I finally tired of cutting and fitting pieces, maneuvering the stitching in all directions. I decided there should be an easier way. To secure practically the same results, I invented my own way of putting pieces together. It's a breeze—even for amateurs.

Although we are using the word "quilt," I make these covers more particularly for spreads. You can make them in all cotton or of materials of approximately the same weight; or, you can make them in a combination of silk and rayon. I also make pillows and other items using this same idea.

The size of the square block can be whatever you want—probably from 8″ to 12″ when finished. Allow at least 1″ for seams.

Cut or tear pieces of old sheeting (or other suitable material for backing) to the size upon which you have decided.

Sort out, or roughly cut, scraps of material a couple of inches longer than your block and from 2″ to 4″ wide or more. Press a stack of these pieces, together with the backing squares. If you spray starch the backing squares, there will be more body to work on.

Use three scraps of material to each square, in random widths. If possible, do the zigzag sewing on the straight of the goods. You can use heavy-duty thread or mercerized—black, white, or a color of your choice. Lay one piece of material to cover an approximate third of the block. Your materials will always be slightly overlapping the squares. Turn the straight edge of another piece under approximately ¼″. Lay this turned-under edge over the raw edge of the first piece of material lying on the backing square. Zigzag with the widest setting down the middle of the turned-under edge. If necessary, lift up the second piece to trim away any excess seam material from the first piece. Cut the second piece to the desired slant. Apply and stitch the third piece over the second piece in the same manner.

From the samples shown, you will see the pieces are of varying and random widths, some straight, some slanted. This is why it is so easy to make these blocks—there is no tedious cutting or fitting.

Turn the block over and trim all edges even with the square backing. Press.

The blocks are stitched together in strips, one block placed horizontally, one vertically. On the second strip the vertical fits against the horizontal, the horizontal against the vertical. This gives the crazy-quilt effect.

One reason I care little for many quilt patterns is because the pattern pieces do not match or fit precisely; the blocks are not squared off when put together.

I figured out an easy way to avoid these imperfections. It requires a slight bit of extra time, but the results are so pleasing that you will not mind any added effort.

Sew two blocks together with a ½″ seam (vertical to horizontal). From the seam (using a ruler, tape measure, or piece of sturdy cardboard cut to exact length), measure and mark with a pencil exactly where the next seam will begin on the No. 2 block, and also

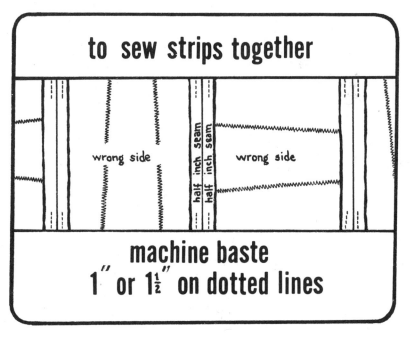

where it will end across the block. For example, if you have started with a 9″ block, after you have made your first ½″ seam, the measurement will be 8″. Concentrate on the inside measurement of the block, instead of on the seam allowance, which is certain to vary slightly. It is virtually impossible merely to lay blocks together and seam with perfect measurement every time. Sew the No. 2 block to the No. 3 block.

It is simpler first to put strips together for the width of the quilt or cover. This eliminates matching longer strips of the length.

Always steam press seams flat on the wrong side before starting to sew strips together. Even though strips of blocks are pinned or basted together, I find there is usually slipping of materials as the sewing progresses, and corners do not match perfectly.

So, with right sides of strips facing, seams pressed flat, match each block exactly where seams come together. Pin if necessary. The strips lie horizontally along the sewing machine. The pressed seams are vertical, or in line with the sewing machine foot. Loosen the tension to a basting stitch and sew about 1″ down the middle of each of the pressed seams. (See sketch.) Continue down the strip, exactly matching seams and machine basting together. Then sew the two strips together. You will find all the pressed seams lying flat, as they should be, and the corners of the blocks matched perfectly. Remove the basting threads and press the long strip seam.

When the entire quilt is put together, zigzag in a continuous line both the width and length seams.

For a cover or a spread I line it with a solid color. The lining should be the same size as the cover. With right sides together, stitch all around four sides, leaving an adequate opening for turning it inside out. If you like, you can zigzag around the top edge. You can tie blocks with yarn. If the cover is quite large and heavy and the lining is inclined either to sag or pull up, you can tack or blindstitch (on the wrong side) along the first row of blocks to eliminate this.

The pillow shown is made of cotton pieces, 8½″ finished blocks. It has an end zipper. The spread is made of silk and rayon pieces, 7½″ finished blocks. The spread is not yet completed, so the top zigzagging of seams is not all finished.

I have covered chair bottoms and stools with these blocks.

I also made a housecoat using the same idea—merely sewing random strips to a muslin butterfly pattern. It has front and back seams, turn-down collar, and front zipper.

Making blocks in this manner is an easy way to achieve the colorful effect of "crazy quilt," and tedious matching and fitting is eliminated.

Ecology Boxes And Pendants

The word "ecology" pops up everywhere. And just about anything is called an Ecology Kit.

I don't recall seeing round ecology boxes. But why not?

These are made from clear plastic cheese containers approximately 4½" in diameter. These cheese boxes usually contain wedges of various samplings of cheeses.

Some of the dried products that can be used to fill the boxes are: beans (all kinds and colors), popcorn, barley, lentils, whole-grain wheat, small macaroni, rice, split peas.

Insets are made with lids, boxes, cardboard sections. These inside containers, or sections, must be very nearly the actual depth of the box. If they are not deep enough, contents will spill over the sides.

Description of the two kits mounted on an artist's canvas board, spray painted yellow: In one kit, a cardboard box lid 2¼″ x 3″ is placed in the center. In another, a round 2½″-diameter tin lid fits against the edge, giving a "sun and moon" effect.

Fill a lid or box with one of the dried products, shake it around gently so that everything settles and nothing creeps over the side. Put the bottom of the cheese container (the deepest portion) over the filled lid, turn upside down, and position the lid where you want it. The bottom of the cheese container becomes the front of the kit.

Fill the remainder of the space with another product. Using combinations of dark and light products gives interesting contrast. Put on the cheese container lid, hold everything together securely, shake around gently, and add more of the product if necessary to fill the container satisfactorily.

Remove the cheese container lid and put glue around its inside edge. Place the lid back on the box and let glue dry thoroughly.

Cut a narrow strip (the width of the lid) of contact paper or plastic tape and edge the outside of the lid to hide the gluing. You could use a decorative tape or braid.

Description of the three-part kit mounted on a ribbonlike hanger: Three bottle tops (these are from Clorox) are positioned in place after being filled with a product. Then the remainder of the kit is filled as described above. You can mix two products, as in this—macaroni and wheat around the lids that contain rice, lentils, beans.

The middle kit has a round 3¾″-diameter lid of lentils surrounded by barley. The third kit is partitioned with pieces of cardboard. One piece goes across the entire box, off center. Fill the smaller section; push the cardboard against this product. Add the other two pieces of cardboard (about 2¼″ long) that make the two smaller wedges, and fill these. Then fill the remaining section. In this kit there are red kidney beans, pintos, soy beans, and lima beans.

Cover the lid rims as described above with contact paper or decorative tape.

For the hanging band, machine stitch a length of sturdy material about 4" wide, right sides together. Turn, as when making a belt, with the seam in the middle. Press with a steam iron. Run one end of the banding through a ring (plastic or metal) for hanging; fasten by tacking securely on the back side. You might have a discarded belt you could use for mounting the kits, or a heavy ribbon or piece of webbing.

The kits can be glued to the cloth banding (allow it to dry thoroughly), or they may be fastened securely with numerous pieces of masking tape, if the entire back is covered, so that the cloth does not pull out. After the kits are taped in place, put one strip of masking tape or self-adhesive carpet tape (remove only one backing) down the entire strip of banding. Cut off tape around the box curves so it does not show in front.

Kits do not have to be mounted—they can be hung singly if you so desire.

Pendants: The larger pendant shown is approximately 1¾" x 2¼"—a small plastic box. Fill the box with a mixture of dried products. Over the lid clasp at the top of the box glue a jewelry cap, with the prongs bent to fit the shape of the box.

Add a jewelry jump ring through the cap. Put a chain (cap and chain should both be gold or both silver) through the ring, and that's all there is to it.

The smaller pendant is a clear square box approximately 1" on the sides and ½" deep.

It is too wide to accommodate a jewelry cap for hanging. So in the top center punch a small hole with a heated pick or awl. Make a hanging ring from a length of fine wire, insert through the hole in the box, and push apart the ends of the wire to hold in place. You might have an odd bit from a piece of discarded jewelry that could be used for this purpose.

Fill the box with wild bird seed. Glue a little ceramic or plaster bird on the top front edge.

Attach a jewelry jump ring through the wire loop; add a chain.

A little glue can be used to hold the lids of the boxes closed, to avoid snagging them open.

You can easily devise other ways to decorate the box. For instance, spray well with clear plastic a tiny sprig or two of dried weeds or straw flowers (the tiny ones); then glue on the outside front.

These unusual pendants are guaranteed to make people talk—about you!

Second Life For Candle Containers

When the candle has completely burned out in the little glass bowl you bought (perhaps it was scented), what do you do with the container? Don't throw it away. There are many uses to which it can be put.

For one thing, you can give it second life as a candle, undoubtedly far more attractive than it was in its first life.

Clean the bowl thoroughly with hot water and detergent. The sample bowl is 4″ in diameter at its widest portion.

Cut two squares of burlap approximately 4½″. Use any color of your choice. Fringe the four edges by pulling out three or four rows of threads.

One of the squares of burlap will be decorated, working on what you will choose as the wrong side.

Tie together with thread four tiny sprigs of dried weeds, strawflowers, and artificial stamens in pleasing combinations. With stems pointing toward the center of the square, glue the sprigs in place at the center edges of the four sides. Cover the ends of the stems with glued-on bits of dried moss or something like fine steel wool. This is to hide the place where the stems are glued.

Now place this "flowered" square on top of the plain square of burlap, the points of the top square lying over the center edges of the bottom square—which gives the effect of an eight-sided star. Holding the two squares together, push them gently into the glass bowl. After they are in place, glue additional four decorations on the top square at or near the four corners, or in folds as the burlap lies in the glass. Use something like tiny acorns—tops and bottoms can be separated. View the arrangement from all angles before completing it.

Now for the candle: There are small votive candles that can be purchased in containers or separately. If you find one small enough to fit through the opening of your bowl, that is fine. Otherwise, it will be necessary to use a substitute. You might have a whisky shot glass or something similar. The container that holds the candle cannot be plastic, since this melts. It must be glass or ceramic.

I made a container from a small glass jar that held spice, a little less than 1¾″ in diameter. If you have a bottle cutter, you are fortunate. I do not have one, so I cut the glass in an old-fashioned way. Saturate a small length of string in kerosene and tie it around the jar where you want it to break—probably 1¾″ to 2″ from the bottom. Light the string with a match and let it burn out. Immediately pick up the jar by the bottom (don't burn your fingers on the glass at the top that is hot) and hold it under cold running water. The top portion will snap off. The break may not be perfect, but this is no disadvantage. You can sand the rough edges, as a bottle cutter would do, but it is not actually necessary so long as you are careful not to cut yourself on the edges.

If you want to make your own candle wax also, melt stubs of old candles in a tin can. From the small tin piece that held a former wick, cut off each side a bit so the remaining piece will fit inside the container you have just made. Attach a piece of string into the tin for the new wick. This piece of string should be long enough to tie or tape to something such as a pencil that can be suspended across two objects to hold the string wick in place as the wax sets up and hardens. Or, you can drape a long piece of string over a cupboard handle to hold the wick in place. When the candle wax has hardened, cut off the wick.

If you do not care to recycle the glass bowl as another candle holder, you could add an additional arrangement of small flowers to stand in the middle; or, you could use small ceramic figurines—human or animals—for an attractive whatnot piece.

The larger scalloped bowl has the two squares of burlap as described above (a little larger in size), and a girl figurine in the center. This girl was a birthday-cake ornament—she is holding a tiered cake shoulder-high. Small strawflowers, weeds, and leaves (without stems) are glued all over the cake to form a huge floral arrangement. She stands on a white plastic oval. This was not harmonious, so some dried strawlike grass is wrapped around her feet, sticking up at random in the bowl, adding to the attractiveness of the arrangement.

Button Portraits

Who doesn't have odd buttons rolling around in a drawer or a box somewhere? If you have large shank buttons (holes on the underside), you can create some very interesting "portraits" to brighten up a wall or a whatnot shelf. You need flat or slightly rounded buttons, either convex or concave, at least 1″ in diameter or larger. You can even use covered buttons if the cloth is plain.

If you like, you can make facial features for the button heads, but I think the mysterious suggestion of a blank face is more appealing.

Hair is made from yarn (wool or rug), crochet thread, fine steel wool, straw thread, angel hair, old hair nets, discarded wigs, et cetera. Arrange the hair in plaits or braids, bangs, long, short, or piled high. Glue in place.

Hats are made from straw, felt, or other suitable materials, clothlike flower petals, et cetera. They could be crocheted. Decorate with tiny ribbons, threads, feathers, yarns, braids, flowers, small pieces of junk jewelry.

Neck and shoulder pieces are felt, fluffs of lace, embroidery or lace cutouts, braids, or other pieces of trimming.

Add tiny buttons, sequins, beads, or tiny chains to complete the costumes.

The portraits are mounted on plastic or metal lids. Black and gold or elaborate cosmetic jar lids make attractive backgrounds. Plastic pull-out ends from rolls of kitchen foil are suitable. An odd piece of miniature pottery or a plate from a doll's tea set could be used. Mason jar lids are ideal. You can paint a lid if you wish, but this is seldom necessary. The rims can be covered with decorative braid or lace.

Most of the lid plaques can be hung by the rim on a small nail. When the inside of the lid becomes the front of the plaque, use double-stick picture mounts or pieces of double-stick carpet tape for hanging.

Assemble and practice placement of the button and all other parts before gluing them in place. Usually the button goes about mid-center, but not always.

These miniature portrait plaques are attractive arranged in a grouping on the wall. It might be fun to caricature actual family members or friends with hair style, clothing, et cetera.

Instead of hanging on the wall, one of the portraits could decorate the top of a special box or container for trinkets and

keepsakes. Or, it could be displayed on a small easel after mounting it on a suitable background.

Here are the descriptions of the sample portrait plaques: Greenish-yellow metal lid, with gold braid glued around the rim. Face: yellow cloth-covered button. Hair: black yarn, bobbed, and with bangs. Hat: orange velvetlike flower petal, with green feather decoration. Neckpiece: green felt with red sequin.

Mason jar lid, with white coating exposed as background. Face: black button. Hair: red yarn, bangs. Hat: piece of thin yellow foam rubber tied in a turban. Neckpiece: blue lace tied in a bow.

A pink background plaque was the bottom from a pot that held a growing hyacinth. Black yarn is woven in and out of the drainage holes. Face: dark tan button. Hair: black yarn laid flat and piled high, with small clothlike pink flower and banding. Collar: red felt. Shoulder piece: blue felt with two red sequins.

Brown background is a piece of pottery resembling a lid. Face: turquoise cloth-covered button. Hair: fine steel wool (grey), with orange buttonlike piece of jewelry for decoration. Neckpiece: folded together piece of orange silk material, with fine gold chain hanging from neck.

Silver lid banded in red braid, with red star sequins glued above head. Face: white pearl-like button. Hair: red yarn braided and tied with one strand of black yarn, bangs. Neckpiece: black felt, with white heart-shaped decoration.

Crazy Helping Hand

You wouldn't dare say you've never lost a glove! And don't you just *hate* to throw away that perfectly good one tucked in the drawer, always getting in the way?

Why not put that lone glove to good use and have a crazy helping hand? It's a hand that will hold a lot of clutter for you, or for a friend, who would be certain to welcome it as a thoughtful gift.

The glove in the photograph is filled with dry papier-mâché (no mixing with water). I know you could use sawdust for filling,

but frankly this might make the hand a bit heavy. If you have some sawdust, you could try it.

Use something thin and narrow, such as a paintbrush handle, a drink muddler, or a pencil, to poke the stuffing down into the fingers of the glove. The fingers should be full, but not actually hard and stiff. They can be moved into different positions, even tacked down to hold a certain shape, if you like.

You need two Mason jar lids and one ring. One lid can be a different kind, but it should be approximately the same size as the Mason lid.

When the glove is fully stuffed, push a lid down on top of the filling, pull up the edges of the glove over the lid and tack securely back and forth across the lid. (If your glove is a long one, cut it off to make a shortie.)

Glue a lid inside the Mason jar rim; then glue the stuffed hand on this lid.

Trim the edge of the rim with braid or ribbon.

You can ask this hand to hold all manner of things—rings, pins, clips, or you can stick in needles, straight pins, safety pins. Hang a necklace here and there. The inside of the hand can also be utilized.

Select one of the fingers to hang the car keys on—then you'll always know where they are.

This is really an ornamental conversation piece, besides being handily useful.

Dictabelt Hat And Belt

I don't know where I got the crazy idea to try to crochet with Dictabelts, but I got it. Here is the result of recycling some Dictabelts my granddaughter gave me. I was supposed to do *something* with them.

Beginning at the edge where the printing is (cut away the printing), cut the Dictabelt in one continuous piece, approximately ⅛″ wide. I'll admit that my cutting was rather erratic and that widths varied. However, this doesn't really matter, as these discrepancies are more or less concealed in the crocheting, and give a textured look as with some yarns. The Dictabelts *can* be cut very meticulously, if you take the time and have the patience. I wanted to get on with it!

Crochet only one Dictabelt length at a time. This enables you to keep straightening out the "thread" as you progress. Use the double crochet stitch.

If you can't crochet without a pattern, more or less fitting as you go, use printed directions for the type hat you select. I used a Size I plastic crochet hook. Adapt the size hook to the plastic and the way you crochet.

The plastic will have a tendency to slip from your hook, but a little experimenting will enable you to get the knack of holding your hook and keeping the plastic where it belongs.

As you are about to finish a length of plastic, tie on another length in a hard knot. Cut off any excess and conceal the knot on the wrong side.

The belt was not made to match the hat, although a hat and a belt could be worn as a set if the colors are the same. The hat in the sample is blue; the belt is red, crocheted with red yarn.

For a belt, fold a Dictabelt in half the long way, creasing it evenly. Cut carefully on the crease. You now have two equal pieces. Prepare enough belt pieces to go around your waist, allowing 2″ to 3″ for overlapping in front. However, before you determine the length, read further for alternate ways of fastening or closing the belt.

You will need an eyelet punch to make holes in the Dictabelts. Start at one corner and space the holes evenly, along both long sides of the belt pieces. Be certain you do not make the holes too close to the edge, or the hole will tear apart. It is well to practice first on a piece of Dictabelt, making the holes, as well as crocheting several stitches to determine that spacing is correct.

With yarn (use the same color or a contrasting color) begin at one end and make two single crochet stitches in the first hole, then one single crochet in each successive hole, except the last hole in the piece. Make two single crochets in the last hole, the same as the first hole. Hold another belt piece snugly against your work and make two single crochets in its first hole. Then make one stitch in each successive hole, except make two in the last hole. Continue crocheting pieces together on one side until the belt is the correct length. Then make holes in the end edge of the last piece and crochet here also, putting three single crochet stitches in each corner hole. Crochet the other side of the belt in the same manner as the first side. Finish the other end piece at the beginning as described, with three stitches in each corner hole. The slits between the belt pieces are not fastened together.

After crocheting around the entire belt, slip stitch into the first row and make a second row of single crochets. If you find the second row is inclined to pull a bit and be too tight, add an extra stitch now and then by making two stitches in one stitch of the first row. This will not be noticeable and will keep your work flat.

There are several ways to close the belt in front. Lap over the ends and pin diagonally (through the crocheted edges, not the belting) with a large jewelry-type safety pin. These pins are sometimes used for lapped skirt closures. This is shown in the photograph. Or, you can fasten the belt with four snaps, sewing the snaps into the crocheted edges. You can also finish the ends of the belt with eyelets (crocheting them before breaking off your yarn), and lace a crocheted cording through the eyelets, tying it in front.

Dolly's Clothes Hangers

Supposedly women never have anything to wear! Perhaps this is true also of our youngsters' dolls. But a more urgent lack seems to be where to hang dolly's clothes. There are never enough hangers. So let's make some.

Pipe cleaners are easily adaptable. Make a small hook at the end of a pipe cleaner; leave a straight length of approximately 1"; then bend the pipe cleaner to the right and make a loop about 1¾" long. Going on top of where you started the loop, cross over to the left side and make a second loop. When the pipe cleaner is brought back to the center of the hanger, making the second loop, you will

have a remaining length a little less than 3″. Use this portion to wrap, knotlike, around the center, tucking or pushing in the end of the wire.

Add a tiny flower or ribbon bow for decoration, if you like.

For larger-sized hangers, use four pipe cleaners. Lightly twist two pipe cleaners together. Make the hanging hook and one arm loop with these, leaving a small length in the center for twisting together. Use two more pipe cleaners twisted together to form the other side loop and knot both pieces together in the middle. Cut off any excess. Using two sets of pipe cleaners in this manner, the hanger can be formed to any desired size.

Other practical hangers can be fashioned from plastic lids. Use the size lid that would be adaptable to the size of the clothes it will hold.

Cut off a portion of the lid—something less than half of it. You can use pinking shears if you have them. Clip off little rim portions at the ends that seem to protrude (this is the back side of the hanger).

You can cut out the inside portion of the lid and use the rim only. You can also shape the inside portion of the hanger with a circular cut.

For hanging skirts, slacks, et cetera, use small safety pins to fasten them through holes you have punched in the hanger with an eyelet or paper punch. You can punch two holes, or more than two, across the bottom of the hanger, to accommodate different-sized skirts.

Using a razor blade and metal ruler, you can cut a straight bar of plastic across the bottom of the hanger, ending at the rims. Clip out the remaining circular part of the lid with a scissors.

With a large needle, awl, or pick, punch a hole in the center of the outside rim at the center top. Push through a length of pipe cleaner formed in a hook. Twist or bend a small portion of the wire on the underside of the rim to anchor the hook in the hanger.

Candle In A Bottle

Sticking a candle in a bottle isn't anything new. What I have in mind is demonstrating how a simple thing such as this project can be individualized by your own creative abilities.

The bottles I selected held wine. They are an unusual shape for a wine bottle. So keep this in mind and, if possible, choose a bottle that is a bit different or out of the ordinary.

The identical bottles have been decorated in two entirely different ways. The one with the short candle is suitable for patios and picnics—occasions that are simple and casual. It is filled with wild bird seed. The candle is placed in a yellow plastic lid, which sits on top of the bottle's contents. The lid to the bottle was not needed.

The other bottle is filled with a mixture of loose beads. A holder for the candle is glued to the snap-on lid. This bottle top is fancy and gold—originally from a cosmetic product. It has a taller, inner circle which holds the candle perfectly. Around the container, on top of the lid, decorative braid is glued, as well as around the edge of the rim. This latter braid has bead dangles attached.

The bottles have a raised circular identification crest. A discarded earring was shaped and glued to this. Add a tall candle and use this candlestick for a more festive occasion.

It is said that a basic costume can be dressed up or dressed down. The same is true with a basic item for decorative or utilitarian use in the home, as these candles demonstrate.

Wallpaper Patchwork

Patches, patches, and more patches! Almost anything nowadays is made with patchwork. Instead of cloth, try using wallpaper to recycle an old box, dish, or tray, to give it renewed life.

If you don't have scraps of wallpaper stashed away in the attic or basement, you can probably secure a sample book from a friendly dealer. I have been given numerous books. Or, ask among your friends for scraps and tail-end pieces.

Instead of using a scissors to cut the paper, tear it. After a trial or two, you will discover in which direction to tear the paper so that the top piece shows at the edge, and not the white underside. Use the most colorful portions of a piece of wallpaper, discarding the plain parts.

The taller, squarish box in the photograph is covered with small squares, torn about 1″. Adapt the pattern or design you create to whatever it is you are covering. Slightly overlap edges of each piece and glue in place. At the bottom of a box, or at the lid, carry the piece over the edge or side, so that no portions of the box remain uncovered. The inside of the box and the lid may be covered in the same manner as the outside of the box. Or, you can use one piece for the inside of the lid and straight strips for the insides of the box. In some places, as where you are working with a continuous hinge across a box, use a piece of paper with three torn sides and one straight-cut edge to place against the hinge.

WHIMSICAL PULL-OFF TOPPERS

For the sides of a box, center a piece at each corner and then work toward the center. Carry the top edge down inside the box. Clip away excess paper at corners.

Keep a damp cloth handy to wipe off excess glue and to keep your fingers from becoming sticky.

To cover the top of the cigar box, it is fitted with tapering pieces from the outside to the center; then an oblong piece is fitted in the middle.

After a box is completely patterned, check all edges to see that they are affixed securely. Then give it numerous coats of clear plastic spray for protection and durability.

The inside bottom of a box may be fitted with felt or velvet. Also, cut a piece of felt for the outside bottom. Glue in place.

If you desire, affix a small lock or a handle from a piece of discarded jewelry. Holes can be punched in cardboard, drilled in metal or wood, to hold a jewelry jump ring, to which a dangle handle can be attached.

Whimsical Pull-Off Toppers

I hope you'll like meeting these crazy guys and gals made of pull-off tops from a well-known beverage sold in brown 7-ounce bottles. They are called Chug a Mugs. The bottle neck and cap lid are wider than ordinary—approximately 1½". When I saw one of these pulled-off caps on a table, suddenly whimsical characters began to take shape in my mind.

These can be made to hold place cards, as "GILMORE," shown in the photograph. Or, they're cute enough just to look at! They make clever party favors.

You might have access to similar pull-offs which I have never seen. The idea is to adapt your own creativeness to what you have or can secure.

In the front of the cap where it separates, a bit of the rubber seal sticks out when the cap is pulled off. Remove these bits of rubber with a sharp, pointed knife.

One of the unpainted caps has been cut away to look more like feet than just a base. If you want to do this, cut and shape before

painting. Actually, the uncut base of "GILMORE" looks like over-size feet as it is.

Paint the cap a color of your choice. Let it dry thoroughly.

All of the characters have small movable eyes, obtainable at craft and hobby shops. These are affixed to three different types of faces, described separately.

Character holding pumpkin: The pull-off top is spray painted flat black. The eyes are glued to invisible nylon thread, each thread tied at the top of the ring. With a thread that is not this stiff, the eyes would hang loosely, but because they were inclined to turn and remain at improper angles, a bit of invisible tape holds the two eyes in place. His hair is wisps of red yarn, glued to the top ring both front and back. The bow tie is green *soutache* braid. Arms are black pipe cleaner, glued to the back of the figure. He is holding a "pumpkin," which is actually a miniature plastic orange. Sprigs of plastic greenery are tucked into the base.

The girl in the center is painted red, has red pipe-cleaner arms, and a piece of red braid around her neck. Her movable eyes and red felt mouth are affixed to invisible tape, which was cut in a circle to attach to the back of the ring. Try not to touch the sticky part of the tape, so as to avoid smudges. The hair is blonde wisps from an old wig. Tie little bunches of hair together with thread and glue it in place. Cover the center area where the hair is glued with a bow or other decoration. This bow is black felt. Her long dress is also a piece of black felt, cut as a small gore would be cut. She holds a toothpick stick to keep the geese under control. The geese are glued to the base. A few leaves of plastic greenery are stuck into the base.

The last character has a face of black felt—a small circle glued to the back of the ring. The mouth is a wisp of red braid, the hair is cotton, the big orange bow is velvet. Turquoise pipe-cleaner arms, matching the turquoise paint of the body, holds a place card (glued to the hands).

Canned Hang-Ups

At the last minute you reach into a cluttered drawer for a pair of gloves. Delay and frustration! Matched gloves always seem to get separated.

This irritating situation can easily be remedied by making containers to keep gloves sorted in pairs. The idea works equally well for a towel rack/holder, or for hanging up children's pajamas or sweaters.

Juice cans, or those that held tomato sauce (about 4" long by 2" in diameter) are a satisfactory size to use for gloves. Remove both ends of the can. Use an electric can opener, if possible. At least, be certain that the inside rims are perfectly smooth, snagproof.

The cans are then covered with felt, wallpaper, or contact paper, or any other material of your choice. They could be painted. Each can may be covered or decorated differently, or they may be finished to match. The samples shown are covered with various scraps of contact paper. Use as many cans as are needed to store your supply of gloves.

Here is a hint for using contact paper to cover something such as a can. I find it practical to cut away only about an inch from each end of the backing, leaving the remaining backing intact. This gives a firmer wrapping. Lap the ends of material at the seam of the can. If wallpaper or felt is used for covering, glue the material together at the seam, also along the edges if necessary.

Cut lengths of cord, *soutache* braid, narrow ribbon, or similar suitable material about 12" long. Fasten two cans together with a length of the cording, which is tied in a knot and passed inside a can. Continue fastening all cans together. On the top can, tie a bow in a length of cording for hanging the set. Hang in the closet or in some other suitable place.

This gadget is guaranteed to make both you and your gloves happy!

It is often difficult to find a practical or suitable spot for mounting a towel rack, and a can, requiring very little space, solves the problem.

For this use, a larger can, about 3¼″ in diameter and 3½″ long, is more satisfactory than the smaller juice can.

Before covering or decorating the can, prepare it for mounting. Use a small piece of wood, a broom handle, or something similar, to

insert inside the can so you can hammer holes with a nail. This brace inside the can prevents denting or bending the can. Position a hole about an inch or so from each end of the can, just removed from the seam lap.

Measure the distance between the holes; then mark on the wall or cabinet the same distance where you will put two finishing nails (no heads). Leave approximately ½″ to ¾″ of the nail protruding from the wall.

After the can is covered, as has been described previously, carefully punch through the covering at the two nail holes. Put the holes down over the nails in the wall, and with pliers or screwdriver or a small tool, bend the two nails upward so the can will be firmly affixed without wobble.

Besides hanging this towel holder on the wall or a vertical cabinet, it can be mounted on unused space under the bottom of the cupboard shelves. Use the holder for either hand towel or dish towel—or fix one for each.

This same holder is ideal to be mounted at children's heights for hanging up pajamas or sweaters, which are often flung almost anywhere. Sweaters will not get poked out of shape as they are when they are hung on a nail or a hook. A larger can is more suitable for hanging a sweater.

Personalize the CANtainers, and the kids will love 'em! "MARK'S SWEATER" is gold pre-cut letters affixed to a felt covering.

Cash Cache

Remember the old sugar bowl or jar that was supposed to hold the family savings up on the top shelf of the cupboard? Of course, everyone knew where the cache was stashed.

Here is a clever, sneaky way to hide dollars or pennies that you want to save for an unexpected need or for a rainy day—and no one will be the wiser.

All that is required is a close-fitting lidded container and a smaller jar or can of approximately the same height to sit inside.

The inside container is the "bank." Fill the space between the two containers with an innocent-looking kitchen product, such as dried beans, lentils, or rice, and who would ever know that there is money or small valuables inside?

You might have an old-fashioned glass crock with a clamp-on lid that would be suitable. One container in the photograph is a peanut jar filled with brown rice. Innocent-looking enough?

Instead of putting your secret bank on the cupboard shelf, you could have it right out in the open—even as a decorative touch when it is filled with another product.

The second sample shows a container filled with wild bird seed. To disarm the curious, an artificial bird perches on the lid. If the lid is plastic, and if the bird has a wire attached, you can punch a tiny hole in the lid and push the wire through to be fastened on the underside. Or, the bird can be glued in place.

You can undoubtedly think up a number of unusual ways to disguise your treasure.

But don't forget where you cached the cash!

Planter-Tray Compartments

If you've looked around while strolling through the plant department in a store, you've no doubt noticed the plastic starter pots or trays thrown away by the dozens. You can probably have them for the asking. If you're a flower and plant grower, you accumulate your own.

Some of the little plastic pots are connected to form a tray of six. These make excellent dresser-drawer organizers and handy containers for countless small items that have the knack of hiding, spilling, getting lost. Some trays or boxes you might use are too shallow, but these little pots have depth enough to accommodate a larger quantity and hold things upright.

You can store buttons, lipsticks, small sewing needs while you work at the machine, a string of broken beads, rings, necklaces, et cetera.

Discard any trays that are cracked or smashed. Wash and dry thoroughly. Glue and stack two sets of trays together, which gives them strength and durability. Then the trays can be dressed up with a bit of braid or decorative trim, if you like. A button container could have buttons glued on the sides for identification of colors.

One day I was fooling around with some of these single plastic pots, green in color. I cut small squares of gold paper and affixed them with tape in the bottoms (you could use glue), to show through the holes. You could glue colorful shank buttons over the holes, or discarded pieces of jewelry.

Now make some sort of an arrangement to hang on the wall, which can look very attractive, especially when grouped with various other hangings and/or pictures.

Tape pots together with masking tape down one side each of two pots, to form whatever pattern you have devised. Then put lengths of masking tape across the tops (now the bottom of the arrangement).

You will probably find your creation a bit wobbly, but it will hang straight when placed against the wall. According to the pattern you have put together, you will need perhaps two or three long nails for hanging, placed where the top edges of two or three of the pots will fit over the nails. Slant the nails slightly upward. Too short a nail will not give proper support.

Heavy Jug Book Ends

What can be more frustrating than having to hassle outsize books, struggling to get them to stand upright on a shelf with book ends that are far from heavy enough? I've had lightweight book ends tumble to the floor more than once.

You can solve this problem with a pair of *really* heavy book ends, made from plastic containers that held an antifreeze product. Any similar-type container is satisfactory.

Cut off level at the top of the jug the threaded portion where the cap was screwed on. Fill the container with clean, dry sand; or use

dirt if necessary or small gravel. Tape the hole securely closed with masking tape.

Mix some papier-mâché according to package directions. Cover the entire surface of the jug with a thin coating of papier-mâché, finishing the surface smooth if you like it that way or leaving it more textured if you prefer. Paint highlights show up better on a surface not too smooth.

Decorate the jug (on one side only) in any manner you wish, such as making attractive indentations with various objects. A kitchen drawer usually contains a number of satisfactory implements. Or, you can build up a decorative motif with small objects. The sample book ends have curlicues of heavy twine glued in place, as well as a stylized flower made from plastic curtain rings—six petals and one larger ring for a center. Raised decorations are glued on after the papier-mâché is dry, or they are pushed into it while it is still damp. Sometimes the object still requires a bit of glue to hold securely.

When all the previous steps are finished and dry, give the

[117]

entire container a coating of Gesso, a thick white undercoating which is obtainable at craft, hobby, and paint stores.

When the Gesso is dry (it does not take long), paint the book ends the color or colors of your choice. The sample was spray painted flat black, then dry brushed with Rub 'N Buff red and gold. The red and gold are metallic colors, but need no buffing when a final coating of clear spray is used. Spray the entire surface with clear plastic spray.

Cut a piece of felt and glue on the bottom for furniture protection.

Now those big heavy books won't talk back to you!

Foil Pie-Tin Centerpiece

A flower petal centerpiece (to hold or not to hold a candle) can easily be made from foil pie tins. You will need three 7″ inside diameter and two 4½″ inside diameter pans. Other sizes could be adjusted.

The cutting and assembling of the flower will be explained next, but after the pieces are cut (when you are actually making the centerpiece), they are to be painted before assembling them.

Cut a large pie tin into eight even wedges, leaving uncut a small circle in the center. Trim off the outer turned rim, reserving

the pieces. Cut each of the eight wedges to resemble a heart, with a small indentation in the center of each petal.

The second 7″ pie tin is cut in the same manner. Hold your right forefinger at the indentation in the heart and crimp the foil with the fingers of the left hand. All of the remaining rows of petals are crimped in this same manner. The eight crimped petals of the second pie tin fit inside the first eight petals, which remain flat, or slightly raised on the outside. Stagger each row or set of petals; that is, the center of a petal goes at the separation of petals on the last row.

The eight wedges of the third 7″ pie tin are cut entirely apart and the pointed centers cut off for 1″ or 1½″. This is done so that the third row of petals can be positioned in toward the center. Lay down four petals as for four points of the compass; then place the remaining four petals between these first four petals.

The fourth row of petals is cut from a small pan, prepared in the same manner as was described for the second 7″ pan. The wedges are not cut apart.

The fifth or last row of petals is also a small pan. The wedges remain intact for this one also, but after the outer rim is cut away, an additional ½″ to ⅜″ is cut off, so that this fifth circle of petals is a bit smaller than the previous (fourth) row.

After the cutting is finished and the rows of petals are all ready for assembling, spray paint all pieces on both sides. Gold on the silver is attractive. Black paint (such as is shown in the photographs) gives a wrought-iron appearance. Before the paint is entirely dry, wipe off some of the paint with a soft cloth. The color remains in the grooves and low areas, highlighted by the silver.

When the paint is thoroughly dry, assemble as explained above, gluing each layer with any good white glue or cement.

If you want to use a candle, find a small bottle cap (paint it if necessary) into which your candle will fit. Glue the cap to the center of the flower.

Stamens for the flower are cut from the outer rim pieces. Cut away the heavy outer edge. Then cut a piece of the foil into three or four narrow strips, leaving a connecting piece of about ¼″ at the end. Make from eight to ten or more of these curled silver pieces to place around the candle. It is not necessary to glue them in place.

If you use the centerpiece without a candle, the stamens could be glued in place for a permanent arrangement.

Margarine-Container Flowers

What would we do without margarine containers? They are useful for so many things after fulfilling their original purpose.

I tried melting some of the thinner, more pliable containers—not those made of the rigid, thicker plastic. Out of the oven emerged these lovely flowers in the photograph.

The container should be clean and dry. Printed color decorations or motifs, as well as price markings, can be removed with lemon extract. A bit of detergent and kitchen cleanser also help.

With a sharp scissors cut away the rim of the container. Then make six pie-wedge cuts into the middle, leaving about a ½″ center uncut. Now shape the wedges into petals. As you cut over the hump where the bottom and side join, take a generous slice here, tapering into your first wedge cut. This eliminates the possibility of the petals sticking together because they touch.

At least until you have some practice, melt or soften one flower at a time, as the plastic could become soft and misshapen while you work on another flower. Put the cut, shaped plastic, bottom side down, on a piece of foil on a pie tin. Place in a heated oven, 350° to 375°. You can leave the oven door open so you can watch it if necessary. If your fingers are tender, you might want to wear a pair of old gloves so you don't get burned.

The petals usually curl up quite naturally as the heat takes effect. If they don't, remove the pan from the oven and help the shaping process a bit with a pair of scissors or small long-nosed pliers. Work quickly; don't make dents in the soft plastic. The plastic will harden quickly; after it has, don't try to manipulate it. It can't be done. Put it back in the oven to soften it again if you need to work with it further.

When the flower is shaped and cool, punch two small holes in the center with a heated pick or awl.

Some of the containers are colored. These make beautiful flowers. Some are colored on the outside, white on the inside. These are pretty as the petals curl up to show a second color. The white flowers can be sprayed any color of your choice. You can make a bouquet all one color or of mixed colors.

If you paint the flowers, be sure they are thoroughly dry before proceeding further.

Put the end of a length of floral wire through the holes in the center; twist together on the underneath side. If you have not used covered wire, plain wire can be wrapped with floral tape.

A suitable center is glued in place last. I have made most of the centers different in the samples to show the endless possibilities. However, so many different kinds of centers gives the bouquet a hodgepodge look. Generally speaking, most of the centers should be more or less the same—perhaps different colors of the same material—to tie them all together.

Here are some materials to be used for centers: a discarded earring, three small chenille balls or one larger ball, tiny artificial cloth flowers, tinsel, pot scrubber, small plastic orange, hairlike strands from old hair foundation, coiled paper ribbon, doubled together snips of pipe cleaner, prongs of two plastic forks melted over a candle, a cotton ball.

You might like to add some plastic greenery to your bouquet.

Clown Bank

Most children would like to have a bank of their very own. Most children also love clowns. So why not combine these two and make your favorite youngster a clown bank?

The body, which will hold almost as much as Fort Knox, is one of the newfangled potato chip containers (stacked chips). Cover it with a piece of contact paper. The sample pattern is in reds and oranges. Instead of removing the entire backing of the contact

paper, cut off only about ½″ at the two edges where the overlapping seam will be. This gives greater durability and is easier to apply. Also, cut a circle of contact paper for the container lid.

With a sharp knife or pointed scissors, make a money slot in the back of the body near the seam. If the cutting is not so neat as you would like, cover the slot with a small oblong or oval of felt, in which you have cut a slit. Glue in place.

The top of the container becomes the bottom of the body, and the lid gives access for taking out the money. (Let it get full first!) Red decorative braid is glued around the rim of the lid. It is also put around the other end of the box, but this does not show beneath the collar.

The head is a styrofoam ball, approximately 4″ in diameter. The eyes are movable, obtainable at craft and hobby stores. The nose is a hard round bottle cap from a cosmetic product. It has red freckles, put on with a marking pen. The smiling mouth is cut from red felt. The inside of the lip outlines was colored red with a marking pen, before gluing the mouth in place.

The hair is red rug yarn. After it was glued in a circle around the head, it was unraveled to make it fluffy and kinky. Also include a few straggly bangs. The eyebrows are pieces of the same red rug yarn, glued in place.

A tall bottle cap, about 1″ in diameter, is pushed up into the styrofoam head, by its rim; then the flat bottom of the cap is glued to the top of the body, originally the bottom of the container. After the features have been affixed to the head, determine the proper placement before pushing in the bottle-cap neck. Use glue, if necessary, to hold it securely.

The neck ruffle is a length of blue satin ribbon. Gather it with needle and double thread, pull up the gathers, and tack the ruffle in place at the back of the neck.

The hat is the larger half of a Leggs hosiery container. A party favor—an umbrella over birds in a flower nest—is pushed into a hole in the hat. The hole in the plastic hat is made with a heated pick or awl. The "handle" of the umbrella was long enough to poke into the styrofoam ball, which helped to anchor the hat. Glue the hat in place over the hair. You could use another type of container for a hat; cover it with a scrap of contact paper. A stemmed flower, glued or poked into the hat, would make another appropriate decoration.

Mason Jar Lid Pendants

This is an idea for a pendant or jewelry piece for a larger person, brave enough to wear it. For a smaller person, it might be a bit overpowering.

The pendant is made from a Mason jar lid and ring.

Cover the top of the lid with a circle of velvet, glued smoothly in place. (The sample is purple velvet.) In the center of the velvet-covered lid, attach a bit of discarded jewelry from an old earring or pin. The sample is a tiny gold umbrella with turquoise rhinestones.

Glue the lid inside the rim.

Around the edge of the rim, where the lid is seated, glue gold fringed braid (or other suitable trim), with the fringe to the inside of the circle or center, abutting ends neatly. Also, glue gold braid or trim around the entire side of the rim, abutting ends neatly at the top.

Bend the prongs of a jewelry cap with eyelet, fitting three prongs on the top of the braid-covered rim and one underneath the rim. Glue in place. Attach a jump ring through the eyelet of the jewelry cap, and then an additional jump ring if it is fastened into a link of the neck chain. If the neck chain slips through the jump ring, one jump ring is sufficient.

For another pendant, fit the jar rim over a styrofoam ball (about 2¾″ in diameter) and slice it off just inside the rim. The lid is glued on the back to this flat piece of styrofoam.

Cover the styrofoam ball with a circle of cloth (you might want to match a costume), smoothing out and gluing the edges securely. Push the ball into the rim and glue in place.

Attach a jewelry clasp with eyelet, as was described above; also use a jump ring and a chain.

Paint Applicator Desk Accessory And Easel

Three idle paint applicators or trimmers stared me in the face for a long time; so eventually I had to do something about them.

This paint applicator or trimmer is a contraption consisting of a metal handle affixed to a foam-rubber sponge, topped with rough-textured material. Remove the sponge padding.

The metal handle has a copperlike finish and is very attractive in itself. I glued the 6″-long handle to the back of a small plaque, thus making a perfect standing easel. The plaque was a thin 4″ x 8″ board which I had painted with an odd-looking cat. It is now more attractive than when it was hanging on the wall.

The 4″-long handle was covered with gold and white mottled contact paper on its two flat sides and inside curve. With the handle placed on its side, this inside curve made a perfect receptacle for pen or pencil, or both. Over the contact paper a bit of decoration cut

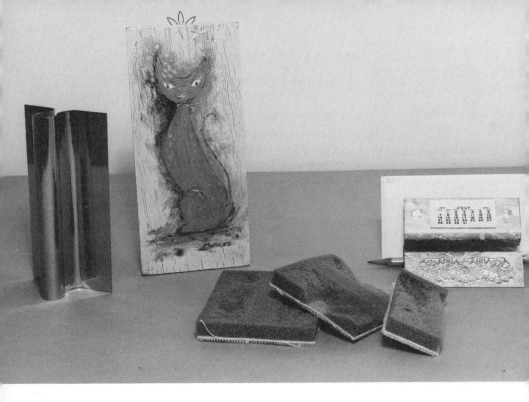

from a gold paper doily was glued at the bottom. A couple of gold dots were glued on the top piece, one on each side of a small calendar (2¼" x ¾", obtainable at stationery stores) glued in the center.

The smallest handle, only 3" long, was turned on its side and attached with Art Metal (you can use epoxy glue) to the back of this 4" handle. This makes a perfect holder for loose notepaper.

Old-Fashioned Clothespins

This might seem like a childish project, but you'd be surprised what an attractive picture holder or standard can be made from a couple of clothespins and a small board.

The 7½" x 4" board I happened to have had rounded contours, which give it a pleasing shape. If you don't have one, for a few cents you could have a small board cut, sand it, and finish it as you wish.

I stained my board and the two clothespins, then glued the clothespins upside down on their heads to the board. This made a perfect holder for the old-fashioned portrait of my mother.

You could use paint instead of stain for a desired color scheme. Instead of a board, you might have a nicely finished plastic lid from a box or something else on that order.

Decorations to suit your fancy could be added to both the board and the pins, but for the portrait I used, I preferred simplicity.

I also used these round-headed clothespins to fancy up an old candy tin. You need a container with straight sides. It is not necessary to even glue the pins in place if the box is not moved a lot. If handled frequently, it might be better to glue the pins.

You can first stain or paint the clothespins, but I use my box to hold sand for starting plant slips and so I like the natural look.

Making round-headed clothespin dolls is not a new idea, but you might be overlooking a clever use for them. They make unusual decorations for gift boxes.

Tiny facial features are made with blue or black and red ink pens. Hair is wisps of old wigs, yarn, cotton, tinsel, old hair nets, et

cetera, glued into place, with your own creative hair style. Hats can be crocheted, made from tiny bottle or tube caps, flower petals, or other things. Arms are shaped from pipe cleaners, wrapped around the clothespin.

Dresses can be made in a real pattern shape, or they can be draped, pinned, and glued in place. Using two thicknesses of material for a skirt gives a good balloon shape and enables the doll to stand alone. Decorations are fancy braids and trims, ribbons, tiny buttons, sequins.

A doll can hold something in her hands. One in the photograph has a bouquet (flower stamens); another cradles a pet lamb. The hippie-type mother on the package clasps a naked baby. Remember that it is that little extra something that makes your own creation special.

String Or Twine Holders

A ball of twine or string is more or less a necessity in every household. The string usually rolls around in a drawer, falls off a shelf, or gets tangled. Here are suggestions for remedying these situations.

A plastic container with a lid, such as one that contained whipped topping (6″ in diameter), makes a fine holder to keep the ball of twine in tow.

Unwanted printing on the lid is easily removed with lemon extract.

The container can be decorated with cutouts, such as flowers or motifs from contact paper.

With a punch or pick, make a hole in the middle of the bottom of the container and pull the end of the string through the hole. Put the lid back on and capture the ball of twine inside.

This container can be kept on a shelf, or it could be hung on the wall. With a paper punch, make a hole in the lid for hanging on a nail. Or, you can cut out a small hole with sharp, pointed scissors.

The blue plastic bottle (with a face) is about 6″ in diameter across the widest part; it is slightly oblong. Any similar container can be used. Cut away all but about 4″ at the bottom of the bottle.

For the face, cut an oblong piece of pink felt to fit the indented bottom. Use pinking shears if you have them. The eyes are movable (obtainable where craft supplies are sold). The mouth and nose are cut from red felt. Glue all features in place. The hair and mustache are bits of fine steel wool. Fine steel wool can be pinched and shaped in almost any fashion. Glue in place. You could make a lady's or a girl's face instead of a man's.

At the right side of the container, a little below midway, screw in a cup hook to hold a small pair of scissors for clipping off the twine. About 1″ up from the bottom edge, punch a hole for the free-hanging end of the twine.

Approximately 1″ from the top back edge, and about 1½″ apart, punch two holes in the center top of the container. Cut a narrow strip of pink felt, poke the ends through the two holes, and make a loop for hanging. The strip of felt can be tied in a knot, or the ends can be stitched together, going inside the container.

[129]

I suppose you could say the third string holder resembles a pig. Whatever kind of animal, it is practical and is large enough to hold two balls of twine. The container is about 6″ in diameter, and has a handle.

Cut away the bottom portion of the jug or bottle, leaving about ¾″ or 1″ below the first indentation at the top.

With a nail, punch two holes in the metal cap, centered about ¾″ apart. Paint the cap red or any other desired color. Let it dry thoroughly.

A flower cut from patterned contact paper makes a decorative "beauty mark" on one cheek. A strip of the same paper goes around the slightly indented back edge.

Screw in a cup hook on one side for holding the scissors.

The twine is pulled out through one hole or two holes (for two balls of twine) in the screw cap.

Attach a length of heavy cord or yarn through the handle, tied in a bow, to make a loop for hanging the holder on a wall.

Styrofoam Packing

Styrofoam is an excellent modern packaging medium. After it has cushioned a product safely to your house, you might give it a second thought before discarding the pieces. The various shapes of styrofoam packing are practically endless; so choices of what to use are many.

Here are a few examples to set you thinking about what you could do with it. I used pieces I happened to have on hand at the moment.

There are paints sold exclusively for styrofoam. Regular enamel paint eats into the material. However, enamel can be used if it is sprayed sparingly and quickly. This gives an unusual effect which is attractive, with the white showing through especially with darker colors. Some of the more compact pieces of styrofoam do take the paint as evenly as other materials.

The small utility tray in the photograph is one of the more solid pieces of styrofoam and took yellow enamel spray very evenly.

Semicircular indentations are used to hold small shell decorations, glued in place. Even letting the tray remain plain and perhaps out of sight in a drawer, the fairly soft material makes a fine receptacle for holding watch, rings, et cetera, when they are not being worn. Sprayed black or brown, a larger tray or compartmented box would be ideal for holding the contents of a man's pockets when they are emptied at night.

The octagon-shaped container or box is sprayed black and has four gold self-sticking seals attached. This can be used for an individual desk-wastebasket. I used mine while I was sewing. Threads, ravelings, and small scraps have a tendency to cling to the material of the box. Needles and pins can also be stuck in the edges while you are sewing. Other sewing accessories aren't always "hiding" when they are placed in this container. The box could also

[131]

hold paper-wrapped candy, unshelled nuts, or any number of other things.

The container holding the flowers is the same shape as the one described above. It is sprayed light blue, looks very dainty. (These two boxes protected a crock of cheese in transit.)

Three strands of gift-type ribbon were braided and pushed into holes on two sides, which made an attractive handle for the "basket." Think of the endless decorations for parties you could make from this material that costs nothing!

I also experimented with making a handle for the basket from clothes-hanger wire. You could sew two narrow ribbons together, or make a casing from other material. This should be about twice the length of the wire handle. Push the wire through the casing and you have a "gathered" handle to attach to a basket. The container could become a small sewing basket, lined with a suitable material. The lining could be attached to the sides of the basket with thumbtacks.

For a larger sewing basket, use a larger container, as shown in the photograph. There are several dividers or partitions in the box, which makes it excellent for separating sewing needs. It is not lined, but this is not necessary as the styrofoam is not hard or rough.

Decorate with fringed braid or other trim. The handle could be any number of things; the sample is a curtain ring. The squeeze-together prongs were pushed into the styrofoam, affixed with a little molding clay that hardens. You might have a knob or a handle from some piece of furniture. You could use a small gold or plastic top from a cosmetic bottle or container.

This sewing basket lid has an indentation along the top. I stuck in a few plastic berrylike flowers for a bit of decoration, but it could have been left plain. I could have made a pincushion (sewed from cloth and stuffed) to fill the indentation. It might be a handy receptacle to hold a pair of scissors to be grabbed in a hurry, without having to open the box.

Another piece of packing was sprayed red and converted to a child's whatnot. The little shelflike protrusions and openings were filled with treasured trinkets. This collects all the miniatures in one attractive place.

The "worm" styrofoam packing can be strung on yarn for unusual window hangings. Begin with a bead on the bottom of the strand; leave approximately 1″ spaces between worms.

Some of the worms have turned-up ends. These can be fashioned into dramatic spire flowers. They are beautiful stark white. Make a tiny eyelet tip on the end of a length of wire. String on a pearl bead for the flower tip. Then push on the worms, using smaller pieces first if the sizes vary. Place every other one of the pieces crossways. After you have strung on the last piece, wrap the balance of the stem with floral tape, pushing it up solidly against the bottom worm. With pliable wire these spires can be curved into graceful shapes.

The sample bouquet is displayed in a cut-glass vase with several branches of an elongated, podlike dried weed. An added touch such as this is what raises your creation above the ordinary.

A boxlike piece of styrofoam, with a raised rectangle in the center, was sprayed with gold to make an unusual reverse shadow box for a small painting.

Now we come to the most dramatic use of all for our scrap styrofoam. Let's make an abstract!

The most suitable foundation or backing is a piece of thin plywood or masonite. You can use an artist's canvas board or a piece of heavy cardboard. The sample is done on an old cardboard print removed from its frame. Canvas board or cardboard may be inclined to warp and should be framed when it is finished.

From your pieces of packing, slice and cut away portions of the thick pieces. A serrated edge is supposed to be ideal for cutting styrofoam, but I found my long butcher knife worked equally well or better. So try whatever kind of cutting implement you have,

including scissors. Cut corners of square or rectangle frames and push the ends into different positions. Shape blunt ends to a point. The existing indentations and raised places will give you ideas. Cut entirely new shapes from the thicker pieces. The worms are excellent for filling in small open spaces.

Fiddle around with an arrangement on the background; then begin to glue the pieces in place. Create as you go. Anything is okay! Large straight pins can be pushed through pieces to help hold them in place while the glue is drying. Do not worry if the pieces do not fit solidly against the background. The wiggles will be taken care of in the next step.

When the glued-on pieces are dry, you are going to cover them with glue-soaked paper. You can use tissue paper, paper towels, napkins, toilet paper—anything that is sufficiently strong to keep from tearing while it is wet. In a bowl make a glue mixture of about 2/3 glue and 1/3 water. Tear off pieces of paper that can be handled easily, dip them in the glue solution, apply first to the bottom or hollowed-out sections. Use a flat paintbrush to push into place in the cracks and crevices. A kitchen gadget, such as a suitably shaped plastic mixer stick, works well also. Do not use anything with a sharp point. Pat the paper with additional glue on the brush if necessary. If you are using a board that will not be framed, cover the sides also. You will have wrinkles, of course. This is what gives dramatic texture and highlights to the completed masterpiece. Overlap edges of the paper; even build up if you want to.

When the paper application is thoroughly dry—it will require at least overnight, perhaps a day or two—apply a coating of Gesso. Gesso is a thick white undercoating, obtainable at craft centers or hardware stores where paint is sold.

The Gesso application hardens quickly. Now paint with tempera colors. Use a short, rather stiff-bristled brush that will reach into cracks and crevices. The sample is done mainly in blues, greens, and yellows, with spots of orange and red. Apply the dark color on the bottom places first. Then build up to the lightest color. Use different brushes for each color, or clean your brush thoroughly in water when changing colors. Blend the edges; overpaint in sections. When the first entire application is finished, there may be white missed spots. You could use an additional color to work into these places. If the paint is hardened in sections, dip your brush in water and blend.

When everything is painted to your satisfaction and is dry, dry brush with gold paint, dragging both the ends and flat sides of your brush over the high areas, but also work some into the lower places. It all depends on how much gold you like. I am usually quite generous with gold highlighting.

After the gold application is completed, spray thoroughly with several coats of clear plastic. Turn the picture while you are spraying, so that all areas are covered. Do this spraying as soon as the gold is applied. Specks and spots of splattered gold are then thoroughly blended.

[135]

I confess that I am not much of an abstract lover (I paint in oils), but I did have fun making this. While gluing on the various pieces, I noticed a small triangular area that was not filled in with a worm. I decided to leave this and see what would happen later. I didn't know what or when.

After the picture was completely finished and framed, I set a miniature vase and flowers in this opening. Then I leaned back and howled with laughter. Now I was a true abstractionist!

Quickly I hunted up other miniatures—a silver angel, a green Buddha with a jewel, a bee, a butterfly, a bird in a nest, a box—and put them in openings in the picture. I sat on the floor and shrieked! My masterpiece was as groovy as some of the "junk" I've seen!

I didn't have the nerve to leave the miniatures in place for the photograph. So here it is, completely naked. Not bad, really!

Meat-Tray Galleries

Here are a couple of projects that should delight our younger friends. Some of them might require a bit of help to complete the projects.

The frames for these displays are meat-tray holders used by groceries and supermarkets—not the solid-type trays, but those with holes or openings for drainage. They can usually be rinsed completely clean.

The ecology gallery would make a splendid Show and Tell project. It is also an excellent way to teach youngsters about various dried beans and legumes, nuts, spices, et cetera.

Use a piece of cardboard, preferably white, for the background. There are two trays—the back one is spray painted black; the front tray is left unfinished. Using two trays gives added depth. The hanger is a piece of black yarn or cord, glued between the two trays, as the two trays are glued together.

One piece or several pieces of each product are positioned and glued in each space. When everything is completely dry, spray the finished gallery with clear plastic.

Some of the products used are: sesame seeds, mustard seeds, whole pepper, white rice, brown rice, grits, caraway seeds, popcorn, different kinds of beans, oats, melon seeds, poppy seeds, and so on. The list is endless. Select what you have on hand, or what you would like to learn about.

A larger meat tray, sprayed silver, houses the picture gallery. Cut the heads a little larger than the openings in the tray. The edges of the pictures are covered by the frame partitions.

This arrangement would be ideal for large families or for school classes. Pictures taken at reunions could be made in suitable sizes to fit the openings. I went through a pile of old snapshots to get enough pictures to fill the thirty-five spaces.

You could make a private gallery of your favorite entertainment or sports stars.

The piece of cardboard on which the pictures are mounted could be taped temporarily to the frame if all pictures are not mounted at one time.

Either one of these galleries would look nice on an easel you have made, described elsewhere in this book. (See page 64.)

Pants-Hanger Candlestick

My daughter gave me a pants or slacks hanger instead of throwing it in the trash. The hanging hook had sprung and would not produce the tension necessary to hold the two wooden pieces together.

Before tossing it in my trash/treasure box, I removed the dangling hook, having no idea what I would ever do with a pants hanger.

The arrangement in the photograph is the end result, after I was suddenly reminded of something sort of Danish modern.

The wood is nicely finished, so there was little to do. With the fixture slightly open, a candle was easily pressed down into the wires, together with a couple of plastic flowers and some greenery.

Perch a bird in the leaves, and presto! you've cheated some gift shop out of a sale.

The bird is yellow, matching the candle. His beak and feet are orange, matching the flowers.

Postage And Pill Boxes

A familiar adage tells us that precious things come in small packages. Well, postage is pretty precious these days; so why not make a pretty container to hold this small necessity—stamps? It is easy for loose stamps to get lost among other things.

Small boxes have a way of accumulating. At least, I hope you don't throw them away. Any small plastic or cardboard box can be adapted to decorating. And what more appropriate decoration than used stamps themselves?

The flat, clear plastic container has one large stamp—1¾" x 1¼"—glued on the top lid. On the back side, four regular-size stamps are overlapped and glued in place.

Another box has only the lid of clear plastic. On this, four regular-size stamps are overlapped and glued in place.

The little open box with the lid raised originally held a ring. The bottom part now becomes the lid. It is fitted with a piece of foam material that held the ring. This can be moistened with water for a perfect stamp sealer.

The long narrow box is cardboard. It held a tube of ointment. The labels on the lid could not be removed satisfactorily, so the top and two sides were covered with contact paper, the box turned upside down. When you are affixing the contact paper, lap the edges under the inside of the lid. The top part, which was formerly the bottom, is covered with six stamps glued in place.

On all of the boxes, the stamp decorations can be coated with clear plastic for longer durability.

If you need a gift for someone who *doesn't* have everything, one of these little boxes filled with postage would STAMP you as a very thoughtful giver.

I don't happen to be a pill taker, but thousands of people are. Perhaps you take vitamins. Here are several perky boxes to carry in your purse.

The flat, clear plastic container has a piece of contact paper fitted both top and bottom.

The box with opaque bottom, plastic top, has a piece of red felt glued on the lid. Two white plastic flowers are glued to the felt. The stem and grass strokes are made with a marking pen.

Another ivory-colored box has "MY PILLS" spelled out with self-sticking silver letters. A flower was fashioned from five O's and the little "mushrooms" are cutouts from the letter C.

The tall container held a popular breath mint. It is decorated with strips of blue metallic self-sticking tape and narrow gold plastic tape.

Personal Bulletin Board

The photograph that was taken many years ago has long since been removed; the free-standing frame is gathering dust in a closet or up in the attic. Why not drag it out of hiding and turn it into a handy personal bulletin board for someone's desk?

Cut a piece of corrugated cardboard (this has a certain amount of depth and will hold thumbtacks) to fit the frame. Then cover the cardboard with burlap or other suitable material. The edges of the material can be fastened down with masking tape; they may be glued in place; or they can be held taut by crisscrossing back and forth with needle and thread.

Cut a piece of heavy brown paper or lightweight cardboard to cover the entire back, including the edges of the frame. Glue in place.

To make a holder for some memo slips, cut off the end of a 4"-wide envelope, a piece about 2½" to 3". Then cut out the front side of this miniature envelope in a semicircular line (similar to a

Kleenex box opening), which enables the pieces of memo paper to be picked out of the envelope. Glue the envelope containing a supply of memo slips to the back of the bulletin board on the cardboard.

Now add the extra convenience of a holder for pen and pencil. I looked in my junk treasures and found two discarded thermometer sheaths. Nurses in the hospital had broken two of my thermometers, and naturally I retrieved the castoff, tubelike sheaths. These holders happened to be grey, the same color as the frame. Perfect! You might have small glass vials or bottles, tin or plastic containers, that would be suitable. You could make a roll from cardboard, tape it together, and cover with contact paper.

This bulletin board is one of my favorite junk treasures.

Papier-Mâché Projects

People from my generation keep telling me they remember working with papier-mâché when they were youngsters in school, tearing pieces of newspaper and dipping in diluted glue. I don't recall doing this; the papier-mâché I am acquainted with is the modern, commercial pulverized type.

The thrilling part is that I have discovered how adaptable this medium is to giving second life to throwaways and castoffs, recycling even broken objects for further use.

Papier-mâché is obtainable where craft and hobby supplies are sold. There are mixing instructions on the package, and I can give you some useful tips.

It is a great convenience to have a turntable on which to work—like the one you may have in your kitchen cabinet or a Lazy Susan. This enables you to turn a project as you work without handling it, which is sometimes awkward and inconvenient, as well as damaging. If you can't produce a turntable, use folded newspaper on which you put a piece of foil or waxed paper, and turn the newspaper to working angles.

Place a bowl of water handy to keep your fingers and tools rinsed. You can smooth papier-mâché with pats of water and a tool, or you can leave it rough-textured. Tools with which to work can be found in the kitchen—knives, butter or sandwich spreaders, short-handled spatulas, spoons, and so on. Use an artist's palette, if you have one.

Decorations and motifs are created by making indentations or adding small objects for build-up. Your own home holds a wealth of these working tools and objects for decoration, such as gadgets for cutting and decorating piecrust, the hole or the point of a bottle opener, fork tines, a meat mallet, and so on. Buttons, string, bits of jewelry, et cetera are glued on or pressed into the wet papier-mâché for build-up.

After an object is covered and decorated, let it dry thoroughly. Sometimes this takes several days if the application is thick or the atmosphere damp. When dry, apply Gesso, a thick white coating substance, also available at craft stores. Gesso dries quickly. Then paint the object a color or colors of your choice, either by brush or spray can. You can use enamel or tempera. Often I combine colors in mottled effects. Several colors are more striking, and they are also an aid in hiding defects. When the paint is dry, dry brush with gold or silver, or use any of the Rub 'N Buff metallic colors. For finishing last, spray lightly with two or three coats of clear plastic or varnish.

Glue a piece of felt to the bottom of the object for furniture protection.

As I like to burn stick incense, I was always faced with the problem of where to burn it—I had no satisfactory holder. The only one I could find in a store was small and allowed ashes to drop on the table. Besides, I thought it was too expensive. So I decided to make my own. This was one of my first papier-mâché projects years ago.

I used an old plate, about 7" in diameter. Anything that is chipped or cracked is suitable—even broken pieces can be glued together first. I glued a spool in the center of the plate and, after

covering it with papier-mâché, shaped it to look like a small stump with cut-off branches. The hole of the spool was kept open, of course, to hold the stick incense. Some free-form flowers were fashioned out of papier-mâché and stuck around the slightly built-up rim.

I used black paint, dry brushed with gold, covered the bottom with felt.

I also finished a small container for matches and a slim jar (probably pickle or olive) for holding incense. Now I have a practical, decorative set. The incense holder is perfect and the entire project cost only a few cents.

No, I didn't drink all the gin originally contained in the big bottle! The bottle was given to me by friends. It is painted with reds, browns, greens, dry brushed in gold, to capture the look of an old Mediterranean piece. The indentations were made with kitchen gadgets. This large container is elegant for holding tall dried grasses and weeds.

The three-piece plaque is made by rolling out papier-mâché between sheets of waxed paper. Make a pattern from paper or cardboard, lay on the papier-mâché, and cut around with a knife, patting the edges into shape. The indentations are made with kitchen gadgets; glued-on string and toothpicks are the build-up. Cut out holes (four in the two top pieces, two only in the bottom piece) for hanging. This is done while you are shaping the papier-mâché.

A discarded gold belt made a perfect hanging adjunct. The large link rings were easily separated and put through the hanging holes.

The large wall plaque is an old board approximately 30" x 12". The vase attachment is part of a plastic bottle, finished at about 9" tall. Something such as this plastic bottle can be affixed with masking tape or tacks to the background. Everything, of course, is covered over with papier-mâché, so almost anything goes for the underparts and foundations not seen.

The vase is decorated with string curlicues and two sizes of buttons pushed into the papier-mâché. If an object does not hold well in wet papier-mâché, use glue also.

Large candlesticks are one of the most fascinating things to make. Use old bottles, jars, lids, saucers, dishes—anything to give the desired shape. Plastic containers can be cut into shapes needed. Under the fluted edges of the finished candlestick in the photograph is a cracked bonbon dish.

Glue and tape all pieces together very thoroughly. This basic step is most important, for you do not want something to fall apart in your hands. (I have had this happen.) Use bits of glue-soaked cotton to hold edges or rims or precarious places that are stubborn to hold.

One candlestick is shown unfinished, ready for a coat of Gesso and final painting. You will note that part of the papier-mâché has been smoothed on and the decorative portion is merely roughly textured applications. You would scarcely recognize this candlestick after it has been completed.

I hope this glimpse into papier-mâché projects will fascinate you and give you the incentive to recycle and beautify many junk pieces. You can practically furnish an entire house with your own creative *objets d'art*.

Sewing Kits

Everybody needs a sewing kit of some kind—a little stowaway container to keep needles and pins, two or three spools of thread, a thimble, a small scissors, a few emergency buttons.

You might think in terms of making a kit for a teen-ager, someone at college, or for a man who has no female around to do his small repairs. A kit is almost a necessity when you are traveling.

A couple of margarine tubs (4″ top diameter) make a handy sewing box to store minimum essentials. These tubs are not the kind made from hard or firm plastic. They are the lighter-weight, more pliable tubs, usually decorated on the sides. Use two tubs that match.

Because the lid is fastened to the box part for opening and closing, this makes the container top-heavy. To overcome this difficulty, weight the bottom with a little plaster of Paris. Mix according to package directions and pour into the tub to about ⅜″ or ½″ depth.

While the plaster of Paris is setting, cut a circle of cardboard that is going to fit on top of the hardened plaster. Cut a circle of cloth slightly larger than the cardboard, approximately ½" to ⅝" all around. On the underside, glue the cloth to the cardboard, stretching the cloth taut so there will be no wrinkles. Or, the cloth can be fastened to the cardboard with needle and double thread. Pull up a running stitch around the edge of the cloth; then crisscross stitches back and forth to secure the cloth.

When the plaster is about to complete hardening, lightly push the covered cardboard circle down into the plaster. If the plaster has hardened, you can secure the circle with glue.

The top lid is fitted with a piece of styrofoam ball, also covered with a piece of the same material as that used for the bottom. You will need only a portion of a 3½" styrofoam ball—you do not want to fill the entire inside of the lid. Cut a circle of cloth to cover the ball section and large enough to draw up with a running thread on the underside. It is difficult to get out all of the wrinkles in the cloth, because the ball is not half-sized, but make your work as neat as possible. Glue the "pincushion" into the lid. Glue a length of thick braid or cord around the outer edge where the pincushion fits into the lid. This is not essential, but makes a nicer finish.

With a large needle, punch two holes about 1" apart in the

bottom container at the top rim. Be sure to go under the rim, but not too far, or the rim edge is apt to crack.

Hold the tubs together to co-ordinate the decorative pattern on the containers; then punch two holes in the top container, the same as in the bottom. The "hinges" are jewelry jump rings. If too much plastic accumulates when you punch the holes, shave this carefully away with a razor blade. Add a latch or handle, if you wish—some fancy bead or shank button or old earring. I used a jewelry cap, pushed together. The small ring on the jewelry cap was pushed through a hole (punched with a large needle) and held in place with a very small jewelry jump ring fastened through it on the underside. A handle is not essential.

Because I thought the top of the lid looked a trifle bare, I glued eight green self-stick patches (round furniture protectors) around the outside edge, with one in the middle.

If you have a label maker, you might print "SEWING BOX" or "SEWING KIT" for the top. You could cut small letters or a decoration from contact paper, in keeping with the motif already on the containers.

Another Sewing Kit: This is easily made from a metal box that packaged adhesive bandages.

The lid is fastened to the box with either two or three tabs. By using a small long-nosed pliers, the tabs can be straightened slightly, the lid removed from the box while working on the project.

Cut a piece of contact paper to fit around the box and lap it in back. The bottom edge will fit into the lower protruding rim. The top edge will begin just below the ridge where the hinges are fastened. Also cut contact paper for the outside top of the lid, rounding corners to fit.

For the inside of the lid, cut a piece of cardboard to fit. Cut a piece of cloth (to match or harmonize with contact paper colors) to cover the cardboard, allowing sufficient edges to turn under on the back, gluing or fastening with needle and thread, as has been previously described.

Put in a small amount of cotton for padding between the cloth and the cardboard.

Glue the pincushion into the lid.

Simple, isn't it? But so handy. This one travels well.

Vases From Cosmetic Containers

Bottles have probably been used for vases for as long as there have been bottles. I will admit there are some pretty-shaped and pretty-colored bottles, but I never liked the screw-top necks, which always gives away the fact that they ARE bottles.

There are containers thrown away all the time which make lovely vases, without that screw-top neck. These are the containers of cosmetics that usually have spray caps, which are fastened on with metal gadgets. Some have labels and lettering that cannot be removed. Speak crossly to them and lay them aside! If you can remove the name and lettering, take a screwdriver (be careful not to let it slip and hurt yourself), pry up that close-fitting metal gadget, and get rid of all the "top" stuff. For pity's sake don't throw it away. Disassemble it and put it with your junk treasures!

Now you have an attractive vase. Fix an eye-pleasing arrangement for practically nothing, and then tell yourself how much you have saved. A look or two in the stores will disclose that similar miniature or small arrangements sell from $1.25 to $2.00 or $3.00. And they don't look a bit more attractive than one you can make yourself. These are thoughtful gifts for someone in the hospital or in a rest home.

If you want to change the clear glass (which most of them are), you can color it with Crystal Glaze, a product obtainable at stores where hobby and craft supplies are sold.

If you use a very small amount of this glaze, the glass will appear naturally tinted. I like blue. An eye dropper can be used for putting the glaze through tiny openings. Keep turning the bottle up and down, rolling it sideways, so the liquid spreads evenly. It will take a little time to do this, but you do not want the glaze to dry in blobs, which will give darker-colored streaks and spots. Actually, this glaze is probably most often used on the outside of the glass, but I like to tint the inside too.

Dried flowers and weeds look attractive in these little vases. Add a small butterfly, bee, or bird to perk up the arrangement.

And when your youngster brings in a dandelion or some other beautiful weed, here is a no-cost container that holds water and will display that terrific bouquet to good advantage!

They're Not Ice Cream Cones!

Sometime ago at a rummage sale I found a couple of heavy cardboard thread cones that are used commercially in sewing factories. When I brought them home I painted them, dry brushed them with gold, then stuck in a number of pushpins (from the bulletin board) at random all over the cones. These made neat holders for rings.

When a friend recently gave me two big sacks of these cones, I was compelled to dream up further ideas.

The cones are made in several different sizes; that is, the heights and circumferences vary, as well as the holes at the tops.

This variation gives combinations you can't get from one size only. The cones are easily spray painted. After painting and dry brushing with gold (if you do this), finish with two or three coats of clear spray.

The first thing I made was a vase for artificial flowers, using four cones—one in the center with the open end up, the other three supporting it in the reverse position.

You will find a sort of "seam" in each cone. Glue these seams together when using more than one; and when you are stacking the cones, line up the seams at one place, which would be the back side.

For the vase, paint four cones a color of your choice—at least one coat. When they are dry, glue two cones together at the seams, in reverse positions. The one with the open end is the center or the vase receptacle. After these first two cones are dried, or at least

substantially so, glue on a third, then a fourth. If you try to glue them all together at the same time, both you and the cones are apt to "come apart at the seams." When the final cone is in place, it is well to put a rubber band around all the cones until the drying is complete.

When the glue is thoroughly dry, give the vase another coat or two of paint. I sprayed mine black and, when it was dry, brushed it with copper Rub 'N Buff. If the inside of a cone will show, naturally this should also be painted.

Another handy receptacle holds pencils, pens, a pair of scissors, et cetera. Cut three cardboard discs and glue to the bottom of three shorter cones (4¾" in height). The fourth cone is 7" tall. The four cones are glued together as described above—the three shorter cones on the outside, the taller cone in the center. These shorter cones have a larger (1" in diameter) hole at the top, and each will hold several pens and/or pencils. The center cone can hold a pair of scissors, a short ruler, and so on.

The receptacle would also hold crochet hooks and knitting needles, paintbrushes, and other things.

The painting and trimming should be completed before assembling cones if more than one color of paint is used. The shorter cones had a dark-blue band at the top, which was difficult to cover with red. So I painted this portion black (with a brush, instead of spraying), as well as a matching band around the top of the center cone. There are indentations here, which determined the width of the banding. After all the paint was dry, I added a narrow band of gold self-sticking decorative tape where the two colors joined.

Next I made candleholders. With a sharp knife slice off a top portion of one of the taller cones, so that it will push down onto the small end of another cone of the same size (upside down). I spray painted this yellow, dry brushed with gold; then clear plastic finished the coating. I found a gold jar lid to fit exactly into the top of the candleholder (originally the bottom of the cone) to hold a votive candle. Just for fun, I put a stretchable orange plastic bracelet, matching the candle, around the bottom.

Another candleholder is made with three of the shorter cones, merely stacked. A gold bottle cap from a cosmetic product exactly fit into the top opening to hold the candle. A portion of the second cone had to be cut off at the top, since the bottle cap was about 1½" deep and pushed against the top of this second, or middle, cone. This candleholder is painted blue, dry brushed with gold.

You will notice I have made no mention of gluing these cones together. This did not seem necessary, but you could use glue if you like.

A larger candleholder was made by stacking nine of the taller, wider cones, each painted a different color, except the first and last, which are black. Then two smaller cones were cut off from the top, which gave a hole circumference that would exactly accommodate a bottle lid to hold the candle. These two cut-off cones were stacked on top of the other nine cones and made a slight separation in the stacking. The first one was painted red; the top or last one was painted black. Instead of multiple colors, you might try two colors only, alternating them.

The wall plaque uses thirty cones. They are painted red, dry brushed with gold. Give the cones at least one coat of paint before assembling, then more coats after the assembling is finished.

Cut a circle of substantial cardboard about 17″ in diameter. Make a hole for hanging about 1″ from the edge. Also punch several holes around the center of the cardboard. These holes are for affixing flowers with pieces of wire, after the cones have been glued in place.

If your circle is not really substantial cardboard, reinforce it with a square of cardboard (as large as you can make it to fit the circle) glued to the back. The holes for wiring, of course, would have to be punched through this double thickness if you use the additional square. I did, as my cardboard circle was not really heavy enough of itself.

Practice placing the thirty cones around the circle. The seams should be on the bottom, against the cardboard. The outside edges of the cones will protrude over the cardboard circle about 1½″. You might want to find a bowl or plate to lay on the cardboard circle, centered, to position the cones around it while you are gluing. Make sure to glue the outside edges of the cones to match—that is, to be even as they make the outside circle edge. The inside tips of the cones will undoubtedly vary some. Watch for the placement of your hanging hole. This should be in the center between two cones, where their sides meet.

When the cones are all glued in place, it will help to weight them with some heavy books while they are drying. But be careful not to get any of them out of place.

When the cones are firmly attached, spray two or three more coats of paint, dry brush with gold, then finish with clear spray.

If you happen to have (or want to buy) one of the larger candle flower circles, this is more easily attached to the plaque than single flowers. Fill in the center of the candle ring with additional flowers, and elsewhere as needed. However, single flowers and/or sprays can be used; this merely requires more wiring. With fine wire attach the flowers through the holes you previously punched and fasten securely in back.

This plaque may hang attractively in an office, a beauty salon, et cetera, as well as in your own home, especially in a hallway or the bedroom.

The candleholders and containers are nice enough to use regularly in your home, but you could really let yourself go to make numbers of them to brighten up your party decoration schemes.

One of the taller cones makes an excellent base for holding artificial flowers. You can stack more than one cone for additional height. Just punch holes with a pick or awl and push in all your short-stemmed leftover flowers to create an outstanding variegated centerpiece or party decoration. This cone can be reused and will not fall apart as will a styrofoam cone which you have to purchase and which usually costs more than you think it should.

Who Sits Where?

Who hasn't had a family reunion or a large get-together where it was utter pandemonium trying to seat everybody satisfactorily?

My daughter's family numbers in the neighborhood of twenty-four when everyone gets together, and it was always a hassle with "you sit here" and "you sit there."

Now we have fun finding our own places, and the place cards are conversation pieces, to say the least. If you make these place cards, someone is going to want to borrow them; but they are continuously reusable, so be sure to get them back. And watch out for a thief at your own table!

[153]

The place cards are made from bottle caps. I used caps from Downy bottles. Two sizes of this product have the same size cap. You can use caps from another product, but they should all match in color, shape, and size—unless, of course, you'd rather mix than match. Scrounge in trash cans at communal or commercial laundries to enlarge your supply. Also, beg from your friends.

If you can use miniatures to fit the personality, habits, or hobby of each guest, that's the best yet. This can turn conversation to hilarity and laughter.

To affix the decorations, use Molding Paste or Art Metal. You can use glue or cement, but often there is only a very small area where the object can be affixed, and glue doesn't always hold quite so well as these other products. Art Metal hardens quickly and holds everything securely. It can also be used as ground areas.

Keep a felt pen of medium width with your place cards (we use blue) and write the name of the guest on the cap. The writing or printing will smudge if it is handled excessively, so the caps should be picked up where there is no ink.

The ink can be removed with soap and water, or a household cleaner, and the cap can be used over and over—unless, of course, your guests talk you out of them and cart them home!

A few of the miniature decorations used and shown in the photograph are horses, fruit, frog, turtle, mushroom/flowers/beetle, bee on hive (large button), duck/umbrella, eggs in skillet, flowers in vase, sleeping baby, dog/fireplug, scissors/sewing, mirror/brush.

[154]

Instead of using just a plain miniature, add a creative something to "special" it up. For instance, put a cuddly blanket around the baby; plant a tree for the squirrel to climb; tie a tiny rag around a sore toe (spray it with clear plastic); cut a pancake turner from tin for the skillet.

Once you get started, the ideas are endless. These place cards may cause more pandemonium than not having any, but give them a try!

Final Throwaways And Castoffs

We have about come to the end of our projects. I'll leave the pages of this book with a few quickies to keep your mind clicking and on the alert for throwaways and castoff treasures.

By now you should be familiar with most of the recycling tricks, so I'll skim lightly over details.

I have a passion for boxes. At a rummage or garage sale I found an elegant brown leather box, lined with cork, a dehumidifier in the lid. It had undoubtedly held cigars or a tobacco product of some kind.

The outside of the lid had been slashed or torn, or something had been ripped off in the center. To all appearances, the box was ruined. But it begged to be taken home.

For some time I pondered what to do about the blemish to give the box another chance in life. Suddenly I remembered a discarded checkbook container or folder made from plastic or simulated leather. The light- and the dark-brown colors matched the box perfectly! I cut out a flower and curlicue motif (already patterned) and glued it in place to cover the damaged portion of the lid. The box had certainly been worth listening to and dragging home!

The little gadget on the wall—a vase for plastic flowers or greenery—is a piece of pipe I saw in a field while I was out walking. It reminded me of a musical note. I brought it home, cleaned it up, painted it black, then dry brushed it with gold paint. Two gold earring baubles were glued on two side holes. I have left out the flowers so that you are able to see the entire shape better.

I also picked up the napkin ring while I was on a walk. It is a piece of round bone, bleached white—too beautiful to pass by. Think how an entire set of such napkin rings would set off your picnic table!

The white plastic bucket held a commercial product—I don't know what. It came from some junk pile somewhere. Cleaned up, trimmed with random flowers cut freehand from scraps of contact paper, it solved the problem of where to keep newspapers in a crowded apartment without adequate closet or storage space. Two rows of blue metallic tape add color for perking up the container.

The flower/butterfly arrangement is mounted on a turquoise plastic platter, the last of a set of dishes. The pink and rose flowers are affixed with florist's clay. The butterflies are exquisite, quite expensive. I picked them up on a half-price sale table at a florist's shop. At the time I had no idea what I'd do with them, but they seem to have flown to the right spot to settle down. They have metal clips for attaching.

How about taking one last guess as to what the boxlike shelf or niche is—or was?

Believe it or not, this was a green plastic container that sat on the bathroom floor to hold a toilet-bowl cleaning brush! Now it can hold a special statue, a vase of flowers, a large candle, a fancy bottle, or other bric-a-brac. I sprayed it flat black to give a wrought-iron appearance, added a sprig of plastic flowers on each side (affixed with fine wire), gave it a candle to hold. It hangs by a nail through one of its many slots.

This converted candle sconce did have a bird perched atop the right-hand edge, but the bird seems to have flown away. Perhaps the photographer frightened it, I don't know.

Do have fun turning junk into treasures!

INDEX

aluminum foil, 48
anodized aluminum easels, 64
apothecary jar glass dome, 42

banks;
    clown, 122
    piggy, 70
basket, hanging, 25
bead mobile, 45
belt sewing kit, 80
bels;
    Dictabelt, 103
    flash cube, 88
    pull tab opener, 20
Bible verse holder, 29
bird cages, 23
bird in a whisk cage, 57
book ends, jug, 116
bookmarks, 37
boquets, button, 32
boquets, tinsel, 78
bottle caps, 43, 47
boxes, 48, 50, 108, 155
boxes, ecology, 93
boxes, pill, 139
boxes, stamp, 139
boxes, trinket, 54
bubbles, plastic packaging, 19, 43
bulletin board, 140
butterflies, metal spout, 38
butterfly mobile, 38
butterfly pendant, 78
button bouquet, 32
button portraits, 98

candle containers, 96
candle in a wine bottle, 106
candles, 97
candlesticks, 144, 151
candlesticks, pants hanger, 138
candlesticks, tape dispenser, 22
cans, cleanser, 35
cantainers, 111
cash cache, 113
centerpiece, pin tin, 118
clothespin dolls, 127
clothespins, 126
clown bank, 122
cosmetic container vase, 148
crochet hook holder, 35
coin purse vase, 87
"cutie pie" jewelry holder, 73

Dictabelt hat and belt, 102
dimensional painting, 15
doll clothes hangers, 104
dolls, clothespin, 127
doorstop, poodle, 48

earring holder or stand, 31
easel type bird cage cup, 25
easels, 64, 125, 127
Easy Crazy Quilt, 89
ecology box pendant, 95
ecology boxes, 93
ecology gallery, meat tray, 136
empty headed pincushion, 85
eyeglass case vase, 86

feed cups from bird cage, 24
flash bulbs, used, 62
flash cube belt and jewelry, 88
floral picture, pull tab opener, 21
flower arrangement under bubble, 20
flower arrangement under glass, 41
flower holder from plastic lemon, 68
flower potting tray, 115
flowers, braid, 57
flowers, button, 32
flowers, flash bulb, 62
flowers, margerine container, 120
flowers, pull tab opener, 21
flowers, stemmed, 22
flowers, styrofoam, 133
flowers, tinsel, 78
flowers, yarn, 57
foil antiquing, 48
foil pin tin centerpiece, 118
fork holders, 28
fringe, yarn, 20

giant paper clips, 51
glass dome displays, 41
glove as a holder, 100
glove keeper, 111

hair curler napkin ring, 17
hamburger press plaques, 59
hanging baskets, 25
hanging cup, 24
hat crocheted from Dictabelt, 102
hats as hanging baskets, 25

iced tea jewelry rack, 57
imitation shell pictures, 70
incense holder, 36, 143
iron on tape silhouettes, 60

jeweled pigtails, 73
jewelry, 33
jewelry, aluminum, 66
jewelry, ecology box, 95
jewelry, flash cube, 89
jewelry, Mason jar lid, 124
jewelry, tin can lid, 76
jewelry holder, "cutie pie", 73
jewelry holder, iced tea rack, 57
jug book ends, 116

keepsake watch, 43
kitchen gadgets, 55

knife holder, 57
knitting needle holder, 36

landscaping spoons, 28
Leggs containers snake, 46
lemons, plastic, 68
light fixture shade basket, 25

magnet needle holder, 30
margerine container flower, 120
margerine container sewing kit, 145
Mason jar lid pendant, 124
match holder, 83
meat grinder vase, 55
meat tray ecology gallery, 136
melted plastic, 34, 120
mobiles, bead, 38
mobiles, butterfly, 45

napkin ring, hair curler, 17
necklaces, 33
needlecraft yarn palette, 29

omlet pan plaque, 59

paint applicator easel, 125
painting, dimensional, 15
pants hanger candlesticks, 138
paper clip, giant, 51
paperweights, 44, 69
papier mâché, 142
party favors, 109
patchwork, wallpaper, 108
patchwork pillow, 92
pattern aluminum jewelry, 66
pencil holders, 50, 69, 141
pendant, butterfly, 78
pendant, ecology box, 95
pendant, Mason jar lid, 124
pendant, tin can lid, 76
picture, imitation shell, 70
picture covers, 20
pie tin centerpiece, 118
piggy bank, 70
pill box, 139
pillow, patchwork, 92
pin cushion, empty headed, 85
pipe vase, 155
place cards, 143
planter, 68
plaques, 144, 152
plaques, hamburger press, 59

plaques, omlet pan, 59
plaques for button portraits, 99
plastic lemons, 68
plastic mold displays, 43
poodle doorstop, 48
postage stamp box, 139
postage stamp dispenser, 82
pull tab opener belt, 21
pull tab openers, 20
pull off toppers, 109

quilt, 89

recipe card holder, 28

scissors case, 87
sea creature, 62
sewing basket, 132
sewing basket, margerine container, 45
sewing kit, belt, 80
slide frame yarn keeper, 30
snake charmers, 46
snap on curler napkin rings, 17
spoons, 27
spray can caps, 54
stamp box, 139
stamp dispenser, 82
standing bead mobile, 45
stationary, 60
stemmed flowers, 22
string holder, 128
styrofoam flowers, 133
styrofoam packing, uses for, 130
styrofoam window hangings, 133

tape dispenser candlesticks, 22
thread cone vase, 150
thread cones, 149
tin can lids, 76
tinsel boquet, 78
towel rack from can, 112
trimming wheel earring holder, 31
trinket box, 54

utility tray, 130

vases;
    coin purse, 87
    cosmetic container, 148
    eyeglass case, 86
    meat grinder, 55
    papier mäché, 144
    pipe, 155
    thread cone, 150
    tin can, 36

wallpaper patchwork, 108
wastebasket, 131
watch, keepsake, 43
whisk beater brid cage, 57
wigglers, 46
window hangings, styrofoam, 133

yarn fringe, 20
yarn flowers, 57
yarn holder, slide frame, 30
yarn palette, 29